Claudia Gebhard

Epigenetics in cancer

Claudia Gebhard

Epigenetics in cancer

Genome-wide analysis of aberrant CpG island
methylation in human tumors

Südwestdeutscher Verlag für
Hochschulschriften

Imprint
Any brand names and product names mentioned in this book are subject to trademark, brand or patent protection and are trademarks or registered trademarks of their respective holders. The use of brand names, product names, common names, trade names, product descriptions etc. even without a particular marking in this work is in no way to be construed to mean that such names may be regarded as unrestricted in respect of trademark and brand protection legislation and could thus be used by anyone.

Publisher:
Südwestdeutscher Verlag für Hochschulschriften
is a trademark of
Dodo Books Indian Ocean Ltd., member of the OmniScriptum S.R.L Publishing group
str. A.Russo 15, of. 61, Chisinau-2068, Republic of Moldova Europe
Printed at: see last page
ISBN: 978-3-8381-2070-6

Zugl. / Approved by: Regensburg, Universität, Diss., 2010

Copyright © Claudia Gebhard
Copyright © 2010 Dodo Books Indian Ocean Ltd., member of the OmniScriptum S.R.L Publishing group

Real success
is finding your lifework
in the work that you love.

David McCullough

Table of Contents

1 INTRODUCTION .... - 1 -

  1.1 Characterization of specific tumor types .... - 1 -
    1.1.1   Leukemia .... - 1 -
      1.1.1.1   Normal hematopoiesis and leukemia development .... - 1 -
      1.1.1.2   Acute myeloid leukemia (AML) .... - 2 -
    1.1.2   Colorectal cancer .... - 4 -

  1.2 The concept of epigenetics .... - 5 -
  1.3 DNA methylation .... - 6 -
  1.4 Biological functions and consequences of DNA methylation .... - 7 -
  1.5 Regulation of DNA methylation .... - 8 -
  1.6 Epigenetics and gene regulation .... - 9 -
    1.6.1   Mechanisms of methylation-mediated gene silencing .... - 9 -
    1.6.2   Cooperation between DNA methylation and chromatin modifications .... - 11 -
    1.6.3   The histone code .... - 14 -
      1.6.3.1   Histone acetylation .... - 15 -
      1.6.3.2   Histone methylation .... - 16 -
      1.6.3.3   Recognition of chromatin modifications and the translation of the histone code .... - 16 -
    1.6.4   Non-coding RNA .... - 19 -

  1.7 Epigenetic alterations during tumorigenesis .... - 19 -
    1.7.1   Global hypomethylation .... - 20 -
    1.7.2   Regional hypermethylation .... - 21 -
    1.7.3   Differential DNA methylation patterns in AML and colorectal cancer .... - 23 -
    1.7.4   Differential histone modifications in tumors .... - 23 -
    1.7.5   Therapeutic strategies targeting epigenetic aberrations .... - 26 -

2 RESEARCH OBJECTIVES .... - 28 -

3 MATERIAL AND EQUIPMENT .... - 29 -

  3.1 Equipment .... - 29 -
  3.2 Consumables .... - 30 -
  3.3 Chemicals .... - 31 -
  3.4 Enzymes and kits .... - 31 -
  3.5 Molecular weight standards .... - 32 -
  3.6 Oligonucleotides .... - 32 -
    3.6.1   Sequencing primers .... - 32 -
    3.6.2   Real-time PCR primers for MCIp .... - 33 -

3.6.3	Real-time PCR primers for ChIP-on-chip	- 34 -
3.6.4	Real-time RT-PCR primer	- 35 -
3.6.5	LM-PCR oligonucleotides	- 35 -
3.6.6	Bisulfite amplicon generation (Nested PCR)	- 35 -
3.6.7	MassARRAY QGE	- 36 -
3.6.7.1	Oligonucleotides	- 36 -
3.6.7.2	Competitors	- 36 -
3.6.8	Bisulfite amplicon generation (MassARRAY)	- 37 -
3.7	**Antibodies**	- 37 -
3.8	**Antibiotics**	- 37 -
3.9	**Plasmids**	- 37 -
3.10	***E.coli* strains**	- 37 -
3.11	**Cell lines**	- 38 -
3.12	**Databases and software**	- 38 -
3.13	**Statistical testing**	- 39 -

4 METHODS .... - 40 -

4.1	**General cell culture methods**	- 40 -
4.1.1	Cell line culture conditions and passaging	- 40 -
4.1.2	Culturing of stably transfected *Drosophila* S2 cells and expression of the methyl binding polypeptide MBD-Fc	- 40 -
4.1.3	Assessing cell number and vitality	- 41 -
4.1.4	Freezing and thawing cells	- 42 -
4.1.5	Mycoplasma assay	- 42 -
4.1.6	Isolation of human monocytes	- 42 -
4.2	**General protein biochemical methods**	- 43 -
4.2.1	Purification of the recombinant protein MBD-Fc	- 43 -
4.2.1.1	Dialysis	- 43 -
4.2.1.2	Affinity chromatography	- 43 -
4.2.1.3	Conservation of the purified MBD-Fc	- 44 -
4.2.1.4	Quantification and quality control of MBD-Fc	- 44 -
4.2.2	Discontinuous SDS-PAGE	- 44 -
4.2.3	Western Blot analysis and immunostaining	- 46 -
4.2.4	Coomassie staining of SDS gels	- 47 -
4.3	**General molecular biological methods**	- 48 -
4.3.1	Bacterial culture	- 48 -
4.3.1.1	Bacterial growth medium	- 48 -
4.3.1.2	Transformation of chemically competent *E.coli*	- 48 -

4.3.1.3	Glycerol stock	- 49 -
4.3.1.4	Plasmid isolation from *E.coli*	- 49 -
4.3.2	Molecular cloning	- 49 -
4.3.2.1	PEG precipitation	- 50 -
4.3.2.2	Restriction endonuclease digestion	- 50 -
4.3.2.3	CIAP treatment	- 50 -
4.3.2.4	Gel purification	- 50 -
4.3.2.5	Ligation reaction	- 51 -
4.3.2.6	Sequencing	- 51 -
4.3.3	Preparation and analysis of DNA	- 51 -
4.3.3.1	DNA preparation from normal cells	- 51 -
4.3.3.2	DNA preparation from clinical samples	- 51 -
4.3.3.3	Agarose gel electrophoresis	- 52 -
4.3.3.4	Restriction endonuclease digestion	- 53 -
4.3.3.5	Quantification of DNA	- 53 -
4.3.4	Polymerase chain reaction (PCR)	- 53 -
4.3.4.1	Primer design	- 53 -
4.3.4.2	Standard PCR for cloning or sequencing of gDNA	- 53 -
4.3.4.3	Real-time PCR	- 54 -
4.3.4.4	MassARRAY quantitative gene expression (QGE) analysis	- 56 -
4.3.4.5	Nested PCR for quantitative methylation analysis	- 57 -
4.3.5	Preparation and analysis of RNA	- 58 -
4.3.5.1	Isolation of total RNA	- 58 -
4.3.5.2	Formaldehyde agarose gel	- 58 -
4.3.5.3	Reverse transcription PCR (RT-PCR)	- 59 -
4.3.5.4	Whole genome gene expression	- 60 -
4.3.6	ChIP-on-chip	- 60 -
4.3.6.1	Chromatin immunoprecipitation (ChIP)	- 61 -
4.3.6.2	LM-PCR	- 63 -
4.3.6.3	Labeling and hybridization	- 66 -
4.4	**Analysis of DNA methylation**	**- 66 -**
4.4.1	*In vitro* methylation of DNA	- 66 -
4.4.2	Generation of an *in vitro* partially methylated gene locus	- 67 -
4.4.3	Bisulfite sequencing	- 67 -
4.4.4	Methyl-CpG immunoprecipitation (MCIp)	- 67 -
4.4.4.1	DNA fragmentation	- 68 -
4.4.4.2	Binding MBD2-Fc to beads	- 69 -
4.4.4.3	Enrichment of highly methylated DNA	- 69 -
4.4.5	DNA Microarray handling and analysis	- 70 -
4.4.5.1	Human CpG 12K arrays	- 70 -
4.4.5.2	Human 244K Agilent CpG island microarrays	- 70 -

4.4.6	Quantitative DNA methylation analysis using the MassARRAY system (SEQUENOM)	- 72 -
4.4.6.1	Principle	- 72 -
4.4.6.2	Primer design	- 73 -
4.4.6.3	Bisulfite conversion	- 74 -
4.4.6.4	PCR amplification	- 74 -
4.4.6.5	Shrimp alkaline phosphatase (SAP) treatment	- 75 -
4.4.6.6	Reverse transcription and RNA base-specific cleavage	- 75 -
4.4.6.7	Desalting the cleavage reaction	- 76 -
4.4.6.8	Transfer on SpectroCHIP and acquisition	- 76 -
4.4.6.9	Interpretation of data output and quality control	- 76 -
4.4.6.10	Calculation of EpiTYPER methylation score ratio	- 77 -

4.5 *De novo* motif discovery .... - 78 -

4.5.1	Algorithm for *de novo* motif finding	- 78 -
4.5.2	ChIP-on-chip peak calling and motif annotation	- 78 -

5 RESULTS .... - 80 -

5.1 Detection of methylated DNA by methyl-CpG immunoprecipitation (MCIp) .... - 80 -

5.1.1	Detection of *in vitro* methylated DNA promoter fragments	- 82 -
5.1.2	Detection of methylated genomic DNA fragments	- 83 -
5.1.2.1	Combination of MCIp and real-time PCR to detect the methylation status of specific CpG island promoters	- 83 -
5.1.2.2	Sensitivity and linearity of the MCIp approach	- 86 -

5.2 Combination of MCIp and 12K CpG island microarray analysis .... - 89 -

5.2.1	Experimental validation of microarray results	- 94 -
5.2.2	Global comparison of CpG island methylation and mRNA expression	- 98 -
5.2.3	Aberrant hypermethylation in patients with acute myeloid leukemia	- 99 -

5.3 Global profiling of cancer-associated CpG island hypermethylation using MCIp combined to 244K CpG island arrays .... - 101 -

5.3.1	Establishment of a new microarray platform	- 101 -
5.3.2	Comprehensive validation of genome-wide CpG island methylation profiles for two human leukemia cell lines	- 105 -
5.3.3	Genome-wide hypermethylation profiling in AML and patients with colorectal carcinoma	- 110 -
5.3.4	Confirmation by MassARRAY (EpiTYPER) data	- 114 -

5.4 General transcription factor binding at CpG islands in normal cells correlates with resistance to *de novo* methylation in cancer .... - 118 -

5.4.1	Basic properties of hypermethylated CpG islands	- 119 -
5.4.2	Defining CpG island regions	- 120 -
5.4.3	Strategies for *de novo* motif discovery	- 122 -

5.4.4	Sequence motifs associate with CpG island regions that remain unmethylated or become hypermethylated in cancer	- 124 -
5.4.5	Sequence motifs and transcription factor binding in normal cells correlate with CpG methylation status in leukemia	- 130 -
5.4.6	Properties of CpG island-associated genes in conjunction with CpG island methylation status and transcription factor binding	- 136 -

6 DISCUSSION & PERSPECTIVES .... - 139 -

6.1	MCIp in comparison with existing methods	- 139 -
6.2	Hypermethylated genes in leukemia cell lines and primary tumor samples	- 143 -
6.3	Towards relevant disease markers for AML	- 147 -
6.4	Establishing DNA methylation patterns through *cis*-acting sequences and combinatorial transcription factor binding	- 150 -
6.5	Perspectives	- 156 -

7 SUMMARY .... - 158 -
8 ZUSAMMENFASSUNG .... - 160 -
9 REFERENCES .... - 162 -
10 ABBREVIATIONS .... - 172 -
11 PUBLICATIONS .... - 175 -
12 ACKNOWLEDGEMENT .... - 176 -

List of Figures

Figure 1-1 Schematic representation of the hematopoiesis .... - 1 -
Figure 1-2 Schematic representation of the biochemical pathways for cytosine methylation,
  demethylation and mutagenesis of cytosine and 5mC .... - 7 -
Figure 1-3 Characteristic domains of methyl-CpG binding proteins .... - 11 -
Figure 1-4 DNA compaction into chromatin .... - 12 -
Figure 1-5 DNA methylation, chromatin structure and recruitment of multiple repressors in a
  hypermethylated CpG island .... - 13 -
Figure 1-6 Post-translational histone modifications .... - 15 -
Figure 1-7 Schematic representation of the function of bromo- and chromodomains .... - 18 -
Figure 1-8 Models for the different mechanisms through which cytosine methylation can
  promote oncogenesis .... - 20 -
Figure 1-9 Histone modification maps and DNA methylation patterns for a typical chromosome
  in normal and cancer cells .... - 24 -
Figure 4-1 Schematic outline of the MassARRAY QGE process .... - 56 -
Figure 4-2 Schematic outline of the EpiTYPER process .... - 73 -
Figure 5-1 Schematic presentation of the methyl-CpG immunoprecipitation approach (MCIp) .... - 81 -
Figure 5-2 Bisulfite sequences of an *in vitro* partially methylated gene locus after MCIp .... - 82 -
Figure 5-3 MCIp detection of CpG methylation in specific CpG island promoters
  using real-time PCR .... - 84 -
Figure 5-4 MCIp detection of CpG island methylation in specific CpG island promoters
  using real-time PCR .... - 85 -
Figure 5-5 Sensitivity (A) and linearity (B) of the MCIp approach .... - 86 -
Figure 5-6 MCIp detection of the *MGMT* locus using quantitative gene expression (QGE) .... - 88 -
Figure 5-7 Sensitivity and linearity of the MCIp approach combined to QGE .... - 89 -
Figure 5-8 Schematic representation of DNA methylation profiling using MCIp and CpG island
  microarrays .... - 90 -
Figure 5-9 Validation of CpG island microarray results by MCIp and real-time PCR .... - 95 -
Figure 5-10 Real-time PCR of DNA fragments including transcription start sites .... - 96 -
Figure 5-11 Bisulfite sequencing of six differentially methylated gene loci .... - 97 -
Figure 5-12 Derepression of hypermethylated target genes by decitabine .... - 98 -
Figure 5-13 Methylation profiles of AML patients .... - 100 -
Figure 5-14 Comparison of both hybridization protocols .... - 103 -
Figure 5-15 Major modifications of the MCIp-on-chip protocol in global screening
  for tumor-specific hypermethylation .... - 104 -
Figure 5-16 Examples of microarray results using different hybridization conditions and
  increasing amounts of DNA .... - 105 -
Figure 5-17 Comparative DNA methylation analysis of U937 cells and normal human blood
  monocytes using methyl-CpG immunoprecipitation (MCIp) .... - 106 -

Figure 5-18 Examples for correlation between MCIp and bisulfite data .... - 107 -
Figure 5-19 Correlation of microarray and mass spectrometry data .... - 108 -
Figure 5-20 Methyl-CpG immunoprecipitation and its validation using MALDI-TOF MS .... - 109 -
Figure 5-21 Study design for identifying disease markers for AML .... - 110 -
Figure 5-22 Hierarchical cluster analysis of AML samples in X- and Y-chromosomal genes only - 111 -
Figure 5-23 Hierarchical clustering of tumor samples and one monocyte as well as
 one colon sample .... - 112 -
Figure 5-24 Age-related hypermethylation correlates with developmental genes .... - 113 -
Figure 5-25 Examples of aberrantly methylated CpG islands in AML samples .... - 116 -
Figure 5-26 Examples of abnormal methylation patterns in AML patients .... - 117 -
Figure 5-27 Functional analysis of commonly hypermethylated CpG island regions .... - 119 -
Figure 5-28 Integral hypermethylation values and DNA methylation status
 in CpG island regions .... - 121 -
Figure 5-29 Expression status of genes associated with CpG island regions .... - 122 -
Figure 5-30 Sequence motifs associated with aberrantly DNA methylated (mCpG) and
 commonly unmethylated CpG island regions (CpG) .... - 125 -
Figure 5-31 Motif enrichment in cell lines depending on genomic location .... - 126 -
Figure 5-32 Sequence motifs associated with aberrantly methylated (mCpG) and commonly
 unmethylated CpG island regions (CpG) depending on their genomic location .... - 127 -
Figure 5-33 Distribution of DNA methylation relative to motif distance in monocytes
 and leukemia cell lines .... - 128 -
Figure 5-34 Distribution of DNA methylation relative to motif distance in murine ES cells .... - 129 -
Figure 5-35 Basic analysis of ChIP-on-chip experiments for Sp1, NRF1 and YY1 .... - 131 -
Figure 5-36 Distribution of transcription factor motifs relative to the three motifs for
 NRF1, Sp1 and YY1 at bound sites .... - 132 -
Figure 5-37 Expression status dependent on the binding of general transcription factors .... - 133 -
Figure 5-38 Correlation between transcription factor binding in normal cells and aberrrant
 de novo methylation in leukemia cells .... - 134 -
Figure 5-39 Properties of consensus sequences that are bound or not bound by the
 corresponding transcription factor .... - 135 -
Figure 5-40 Hierarchical clustering of significance values for gene ontology enrichment .... - 138 -
Figure 6-1 A model for DNA methylation protection by the combinatorial action of general
 transcription factors .... - 152 -
Figure 6-2 Transcription factors protect from *de novo* methylation .... - 153 -

List of Tables

Table 1-1	Cytogenetic-based risk stratification	- 4 -
Table 1-2	Genes frequently methylated in acute myeloid leukemia (AML) and colorectal carcinoma	- 23 -
Table 4-1	Elutriation parameter and cell types	- 42 -
Table 4-2	SDS-PAGE stock solutions	- 45 -
Table 4-3	SDS-PAGE gel mixture	- 45 -
Table 4-4	Agarose concentration for different separation ranges	- 52 -
Table 4-5	Reaction parameter for analytical PCR	- 54 -
Table 4-6	Reaction parameter for real-time PCR	- 55 -
Table 4-7	Reaction parameter for nested PCR	- 58 -
Table 4-8	Reaction parameter for 1st LMPCR	- 65 -
Table 4-9	Reaction parameter for 2nd LMPCR	- 66 -
Table 4-10	Reaction parameter for bisulfite conversion	- 75 -
Table 5-1	Hypermethylated gene fragments in myeloid leukemia cell lines	- 91 -

1 Introduction

1.1 Characterization of specific tumor types

The main focus of this work is the epigenetic characterization of two specific tumor forms, in particular acute myeloid leukemia (AML) and colorectal carcinoma. The following sections provide an overview of cancer development as well as genetic and epigenetic features that are associated with the respective tumor form.

1.1.1 Leukemia

1.1.1.1 Normal hematopoiesis and leukemia development

The term hematopoiesis describes the formation of all blood cellular components as represented in Figure 1-1. The cell system is tightly controlled and characterized by a remarkable cellular turnover that constantly regenerates from very few hematopoietic stem cells (HSC) (Steffen et al., 2005).

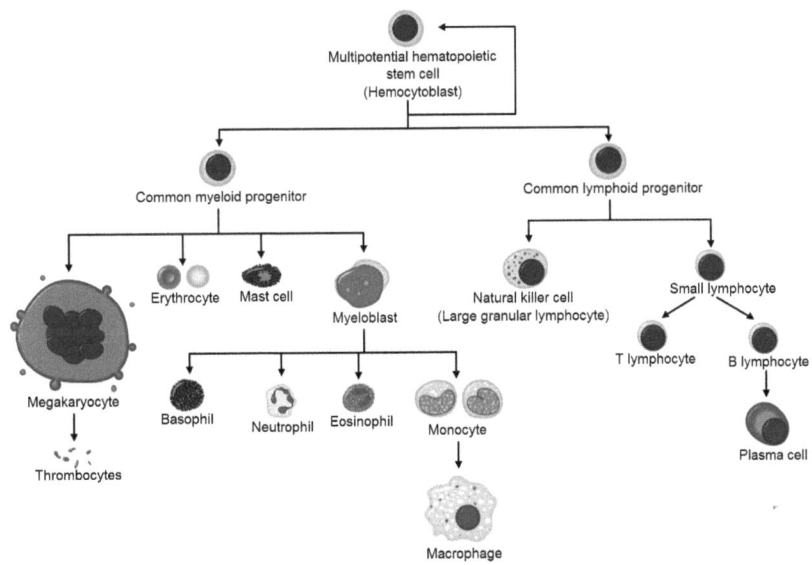

Figure 1-1 Schematic representation of the hematopoiesis
All blood cells develop from pluripotent stem cells. Pluripotent stem cells have a self-renewal capacity and can also differentiate towards either the myeloid or the lymphoid pathway (Wikipedia contributors, 2010).

HSCs reside in the bone marrow and have the capability to give rise to any one of the separate blood cell types. In addition, they are self-renewing and have the potential for asymmetric division. While proliferating, one daughter cell remains as HSC, whereas the other daughter cell develops towards either the myeloid or the lymphoid pathway. Common myeloid progenitors (myeloblasts) differentiate into granulocytes, macrophages, megakaryocytes and erythrocytes whereas T-cells, B-cells and natural killer cells are derived from common lymphoid progenitors (Orkin, 2000). Blood cell precursors progress through a series of stages in the bone marrow before entering the circulating blood stream. While the process of cell division is driven by early and lineage-specific growth factors and their receptors, the decision of differentiation is determined by specific transcription factors that activate lineage-specific genes (Larsson and Karlsson, 2005; Steffen et al., 2005). Because of the high cell division rates of the progenitor cells there is an obviously high probability for mutations which accumulate in stem cells if not recognized by the cellular repair system. Consequently, progenitor cells may lose their ability to differentiate and escape the regulation of proliferation which can lead to the formation of hematopoietic tumors such as leukemia (Steffen et al., 2005). Leukemias can be clinically subdivided into two groups: (A) Acute leukemia which is characterized by the rapid progression and accumulation of malignant cells and is therefore lethal without therapy within several weeks or months. (B) Chronic leukemia typically shows a much slower progression of disease, even if untreated, patients can survive for months or even years. White blood cells for this kind of malignancy are relatively mature but still abnormal. Both groups of leukemia can be further subdivided into lymphocytic and myeloid leukemia depending on their hematopoietic origin. In the present work, acute myeloid leukemia (AML) cell lines or primary AML samples were analyzed.

1.1.1.2 Acute myeloid leukemia (AML)

AML represents a clonal myeloid stem cell disorder that results from genetic and epigenetic alterations. Both, differentiation arrest and excessive proliferation in the immature progenitor pool result in the accumulation of non-functional progenitor cells, termed myeloblasts in the bone marrow and the peripheral blood, where they interfere with the production or the functions of normal blood cells (Jabbour et al., 2006; Shipley and Butera, 2009; Stone et al., 2004). The development of AML has been associated with several risk factors such as age, exposure to viruses, radiation, chemical hazards and previous hematologic diseases or chemotherapy as well as genetic disorders (Deschler and Lubbert, 2006).

Genetic events that are crucial for leukemic transformation comprise alterations in myeloid transcription factors as well as mutations of signal transduction intermediates (Steffen et al., 2005). Specific cytogenetic abnormalities are described in many patients with AML. Cytogenetic events involve inversions, deletions and balanced translocations that often result in the fusion of two genes at the chromosomal breakpoints (Steffen et al., 2005). Abnormal fusion proteins such as AML1-ETO, PLZF-RARa and MLL fusion proteins are expressed and can cause a block of differentiation. Most, if not all of those fusion proteins can recruit corepressors and histone deacetylases, which in turn induce conformational changes of the DNA structure. Consequently, the DNA accessibility for the transcription machinery is impaired leading to the repression of target genes. Another example is the t(15;17) translocation. The encoded PML-RARα fusion protein disrupts the normal response of RARα (retinoic acid receptor α) to retinoic acid. It binds to the retinoic receptor element in the promoter of several myeloid specific genes and inhibits differentiation of the cells (Steffen et al., 2005). Cytogenetic aberrations often have prognostic significance. Translocations such as t(8;21)(q22/q22) and t(15;17)(q22/q12) or inversion inv16(p13;q22), creating the fusion proteins AML1-ETO, PML-RARa and PEBP2βMYH$_{11}$, respectively, are associated with good outcomes after treatment. In contrast, AML patients with a complex karyotype, partial chromosomal deletions (e.g. 5q) or deletion of whole chromosomes (5 and/or 7) are known to respond poorly to treatment (Table 1-1). However, recent studies revealed many genetic abnormalities that escape classical cytogenetic detection (Lowenberg, 2008). Changes in expression levels may be due to small amplifications or deletions as well as point and/or frameshift mutations in the coding region of critical genes. Constitutive activation of signal transduction molecules was observed in tyrosine kinase receptors Flt3, Ras, and Kit (Lowenberg, 2008). For example, thorough sequencing of many mutant alleles from patient samples revealed internal tandem duplications (ITD) of varying lengths in the juxtamembrane region of the Flt3 receptor (Flt3-ITD). In cell line models constitutive autophosphorylation of Flt3-ITD has been shown to facilitate cellular proliferation independently of external growth factors (Steffen et al., 2005). Other somatic mutations have been observed which affect transcription factors playing an important role in lineage-specific differentiation. Examples include PU.1, C/EBPα and GATA-1.

Introduction

Table 1-1 Cytogenetic-based risk stratification (adapted from Appelbaum et al., 2001; Jabbour et al., 2006; Shipley and Butera, 2009)

Risk category	Abnormality
Favorable	t(8;21), t(15;17), inv(16), t(16/16)
Intermediate	Normal karyotype, t(9;11), del(7q), del(9q), del(11q), del(20q), +8, +11, +13, +21, -Y
Unfavorable	Complex karyotype, -5, -7, inv(3)/t(3;3), t(6;9), t(6;11), t(11;19), del(5q)

Recently, considerable progress has been made in our understanding of the genetic processes involved in transforming hematological cells. The increasing numbers of cytogenetic and genetic abnormalities (markers) detected in AML allow for further dissection of AML into molecular subtypes with distinct prognosis. To date, there are two commonly used classification schemata for AML, the French-American-British (FAB) system and the newer World Health Organization (WHO) classification. According to the FAB classification AMLs are categorized into subtypes, M0 through M7, based on the type of cell from which leukemia developed and the degree of maturity of the leukemic cells. The WHO classification is an advancement of the FAB classification and includes more meaningful prognostic information such as morphological, immunophenotypic, genetic and clinical criteria (Bennett et al., 1976; Vardiman et al., 2002). The distinction of specific subtypes of disease with different prognosis enables risk-guided and targeted treatment strategies optimized for each patient (Lowenberg, 2008; Shipley and Butera, 2009; Stone et al., 2004).

However, despite aggressive therapy, only 20-30% of patients enjoy long-term disease-free survival (Shipley and Butera, 2009).

1.1.2 Colorectal cancer

Colorectal cancer describes cancerous growth in the colon, rectum and appendix that represents about 95% of all colon tumors. It constitutes the third most common form of cancer and the third leading cause of cancer-related deaths in the Western world. Colorectal cancer became one of the most frequent malignant diseases in Europe and affects about one million people world-wide each year. The development of this neoplastic disease represents a multistep process in which genetic and epigenetic alterations accumulate and consequently lead to the transformation of normal colonic epithelial cells to colon adenocarcinoma cells (Grady and Carethers, 2008). Genetic abnormalities include hereditary as well as somatic mutations in specific DNA sequences, affecting in

particular DNA replication or DNA repair genes (Ionov et al., 1993). *APC, K-Ras, BRAF and p53* genes (Ades, 2009) are also often mutated leading to uncontrolled cell division. *APC* mutations, for example, play a critical role in the inherited familial adenomatous polyposis (FAP) which represents a predisposition to cancer (Grady and Carethers, 2008).
A key molecular step in the early tumorigenesis process of colon cancer formation is the loss of genomic stability. In colon cancer three forms of genomic instability have been described: microsatellite instability (MSI), chromosome instability (CIN) (gains and losses of chromosomal regions), and chromosomal translocations (Grady and Carethers, 2008).
Colorectal cancer staging describes the depth of penetration, whether it has invaded adjacent organs and whether it has spread to lymph nodes or distant organs and is important for choosing the best method of treatment. The most used staging system is the TNM system of the American Joint Commitee on Cancer (AJCC). "T" describes the degree of invasion of the intestinal wall, "N" the degree of lymphatic node involvement and "M" the degree of metastasis (Greene, 2007). Additionally, the numbers I, II, III, IV describe the tumor progression with higher numbers indicating worse prognosis. Staging of cancer is an important and powerful predictor of survival and treatment methods.
When colorectal cancer is detected at early stages with little spread, it is curable in up to 80% of the cases. The primary treatment is surgical while chemotherapy and/or radiotherapy may be recommended depending on the individual patient's staging and other medical factors.

1.2 The concept of epigenetics

The identity and the developmental potential of a cell are not only defined by its genetic component. The primary DNA sequence is only a foundation for understanding how the genetic program is read. Superimposed upon the DNA sequence (the genetic code) is a second layer of information, called the epigenetic code (Bernstein et al., 2007). The term "epigenetics" was first used by Conrad Waddington to describe "the causal interactions between genes and their products which bring the phenotype into being" (Waddington, 1942). At present "epigenetics" refers to heritable changes in gene expression without a change in DNA sequence (Goldberg et al., 2007). The key modifications conferring epigenetic control are DNA methylation, histone modifications, which interplay with each other, with regulatory proteins and with non-coding RNAs and thus define the chromatin structure of a gene and its transcriptional activity (Delcuve et al., 2009). The present work particularly focuses on DNA methylation, which provides a stable, heritable and critical component of the epigenetic code.

1.3 DNA methylation

The four bases adenine, guanine, cytosine and thymine spell out the primary sequence of DNA. In addition there exists a "fifth base" which is produced by covalent modification of postreplicative DNA. DNA methyltransferases (DNMTs) transfer the methyl group that is provided by S-adenosylmethionine (SAM) to the carbon 5 position of a cytosine residue to form 5'-methylcytosine (5mC) (Figure 1-2) (Herman and Baylin, 2003; Singal and Ginder, 1999). In mammals, this modification is mainly found on cytosines followed by a guanine, the so-called CpG dinucleotides (CpGs). CpG dinucleotides are not equally distributed throughout the mammalian genome and are also vastly underrepresented (Fazzari and Greally, 2004; Ng and Bird, 1999; Razin, 1998). The human genome contains only 5-10% of the CpG dinucleotides compared to what would be statistically predicted, which is probably due to a process of natural selection (Singal and Ginder, 1999). One possible explanation for this distribution is the tendency of methylated cytosines to deaminate. Deamination of cytosine gives rise to uracil which is recognized as foreign by uracil-DNA glycosylases and correctly repaired. In contrast, deamination of mC gives rise to thymine, which is also a naturally occurring genomic base, not be recognized as "misplaced" and therefore prone to mutation and depletion in the genome over time (Fazzari and Greally, 2004). Despite their relative underrepresentation, CpG dinucleotides can be accumulated in small stretches of DNA. GC-rich sequences are present in satellite repeat sequences, middle repetitive rDNA sequences, centromeric repeat sequences and CpG islands (CGI) (Herman and Baylin, 2003; Plass, 2002). CGIs are often defined as regions longer than 500 bp with a GC content of 55% or higher and a ratio of observed versus expected CpG frequency of 0.6 or greater (Gardinergarden and Frommer, 1987; Plass, 2002; Plass and Soloway, 2002; Takai and Jones, 2002), and frequently associate with promoter regions of housekeeping genes as well as up to 40% of tissue-specific genes and are usually unmethylated (Antequera and Bird, 1993).

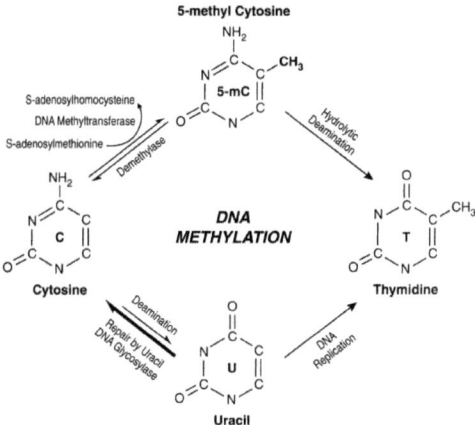

Figure 1-2 Schematic representation of the biochemical pathways for cytosine methylation, demethylation and mutagenesis of cytosine and 5mC
Cytosine can be methylated to form 5-methylcytosine. Deamination of 5-methylcytosine gives rise to thymine, whereas deamination of cytosine gives rise to uracil, which is normally recognized by the uracil-DNA glycosylase (Singal and Ginder, 1999).

1.4 Biological functions and consequences of DNA methylation

About 1% of total DNA bases in human somatic cells constitute 5mC (Ehrlich and Wang, 1981). Nearly 80% of the CpG dinucleotides that are not associated with CpG islands are methylated (Bird, 2002; Herman and Baylin, 2003). Methylation of CpG sites is generally correlated with transcriptional silencing which is thought to prevent the transcription of large and potentially harmful parts of the genome that consist of repeat elements, inserted viral sequences and transposons (Herman and Baylin, 2003). In contrast, the majority of the dinucleotides in CpG islands, especially those associated with gene promoters, are usually unmethylated, whether or not the gene is being transcribed (Herman and Baylin, 2003). An exception to this rule are those islands of genes involved in imprinting and X chromosome inactivation as well as embryonic development and tissue-specific differentiation (Mohn and Schubeler, 2009).

Genomic imprinting is a process of establishing gene expression patterns in a parent-of-origin specific manner (Li et al., 1992). While the vast majority of genes are expressed equally from both parental alleles, some genes are only expressed from one of either alleles due to epigenetic silencing of a specific allele.

Introduction

The inactivation of all but one X chromosome is a mechanism of dosage compensation and is achieved by synergistic expression of *Xist* (X-inactive specific transcript) RNA from the inactivated chromosome, histone deacetylation and methylation (Avner and Heard, 2001).
Controlled DNA methylation is also crucial for gene regulation during embryonic development (Okano et al., 1999). During gametogenesis and embryogenesis dramatic changes in genome-wide patterns of methylation are observed (Kafri et al., 1992; Monk et al., 1987; Reik et al., 2001). Global demethylation after fertilization is followed by waves of *de novo* methylation at the time of implantation. Not all sequences in the genome, however, are demethylated upon fertilization and not all sequences become *de novo* methylated after implantation. These exceptions further emphasize the regional specifity of genomic DNA methylation (Reik et al., 2001; Robertson, 2002).
In mammals, there are at least 200 differentiated cell types, each of them containing the same genome, but using only a small proportion of available genes. Tissue-specific differentiation occurs without changes in DNA sequence (Ohgane et al., 2008). Genome-wide DNA methylation profiles store the "cellular memory" of gene-set activity that governs tissue/cell type feature and is heritable to the next cell generation (Ohgane et al., 2008). However, the extent of tissue-specific methylation profiles throughout the genome is largely unknown and has been the subject of much debate (Walsh and Bestor, 1999; Warnecke and Clark, 1999).

1.5 Regulation of DNA methylation

The establishment of DNA methylation patterns during embryonic development as well as the maintenance and regulation of CpG methylation are not yet fully understood (Ng and Bird, 1999; Razin, 1998; Suzuki et al., 2002). In mammalian cells, three DNA methyltransferases (DNMT) have been identified. DNMT3a and DNMT3b are *de novo* methyltransferases, which are strongly expressed during germ-cell development and early embryogenesis, but at low levels in somatic cells (Klose and Bird, 2006). On the other hand, DNMT1 has a preference for hemimethylated DNA and was therefore assigned to function in maintenance methylation during DNA (Costello and Plass, 2001; Plass and Soloway, 2002). DNMT1 is ubiquitously expressed in somatic tissue and was identified in an enzyme complex together with proliferating cellular antigen (PCNA) located at the replication fork (Costello and Plass, 2001; Plass, 2002). Other components of this protein complex are histone deacetylase 2 (HDAC2) and a DNMT1-associated protein (DMAP1) both mediating transcriptional repression (Plass, 2002). All three enzymes are essential for embryonic development (Costello and Plass, 2001). Mouse embryos lacking both copies of DNMT1 or DNMT3a die before birth, while DNMT3b deletion leads to death a few weeks after birth (Plass, 2002).

DNA methylation is a dynamic but tightly regulated process. Since certain developmental events also involve erasure of the methylation pattern, an enzyme with demethylating activity has been suggested and debated (Plass, 2002). Three main biochemical mechanisms have been proposed that may result in active demethylation: removal of the methyl group, excision of the methylated base or excision of the methylated nucleotide (Bhattacharya et al., 1999; Gehring et al., 2009; Zhu et al., 2000). As opposed to plants, in mammals no specific demethylase has been identified so far, but enzymes involved in DNA repair are potential factors in the DNA demethylation process. It was assumed that glycosylases and endonucleases could cleave and relieve 5mC from DNA followed by repair of the affected site (Jost et al., 1995). Furthermore, base excision repair enzymes, glycosylases and DNMT3a/b have been found within the *pS2* gene promoter. In this model system dynamic CpG demethylation and CpG remethylation processes are inherent to transcriptional cycling of the *pS2* gene, implying a role of DNMTs in demethylation events beside DNA repair enzymes (Metivier et al., 2008). Another, alternative explanation for DNA demethylation could include DNA replication in the absence of maintenance methylation, resulting in passive demethylation (Costello and Plass, 2001).

1.6 Epigenetics and gene regulation

1.6.1 Mechanisms of methylation-mediated gene silencing

CpG methylation, the most abundant epigenetic modification in vertebrate genomes, plays an essential part in the control of gene expression. DNA methylation is normally linked with stable transcriptional silencing of associated genes and much effort has been invested into studying the mechanisms that underpin this relationship. Two main models have been proposed to explain how transcriptional repression may be achieved. (A) The methyl group points into the major groove of the DNA and the space occupied can directly block the binding of transcription factors. Several transcription factors, including c-Myc/Myn, CREB/ATF, E2F and NFκB as well as the regulatory protein CTCF, recognize sequences that contain CpG residues and binding to each has been shown to be inhibited by

Introduction

methylation (Allis et al., 2007) (Bell et al., 1999; Singal and Ginder, 1999). (B) The second mechanism involves proteins that detect methylated DNA through methyl-CpG binding domains (MBDs) (Plass, 2002). MeCP1 and MeCP2 were the first two methyl-CpG binding activities to be described (Esteller, 2005). While MeCP1 was originally identified as a large multi-protein complex, MeCP2 is a single polypeptide with an affinity for single methylated CpGs (Esteller, 2005). Characterization of MeCP2 led to to the identification of two domains, a methyl-CpG binding domain (MBD) and a transcriptional repression domain (TRD) (Esteller, 2005). Database searches identified additional proteins with DNA binding motifs related to that of MeCP2 and designated the MBD family comprising MeCP2, MBD1, MBD2, MBD3 and MBD4 (Figure 1-3) (Allis et al., 2007; Wolffe et al., 1999), with MBD2 being the DNA binding component of MeCP1 complex. Three of the MBD proteins, namely MBD1, MBD2 and MeCP2, have been implicated in methylation–dependent repression of transcription (Bird and Wolffe, 1999) (Allis et al., 2007). Another methyl-DNA binding repressor called Kaiso exists, which lacks the MBD, but recognizes methylated DNA through zinc-finger domains (Klose and Bird, 2006). The proteins have different affinities towards 5mC from MBD3 showing very little affinity to MBD2 that can bind to a single CpG residue (Ballestar and Wolffe, 2001; Fraga et al., 2003). Recently, it has been shown that the MBD of MeCP2 recognizes the hydration of methylated DNA rather than 5mC itself (Ho et al., 2008). Knowledge of the target site of the MBD domains is a prerequisite for understanding its biological role. Klose et al. could show that, despite of their overlapping DNA sequence specifity, each methyl-CpG binding protein is targeted independently in the genome (Klose et al., 2005). MeCP2 strongly prefers mCpG sites flanked by a run of AT-rich DNA, whereas MBD1 has an additional DNA-binding domain specific for non-methylated CpG (Klose et al., 2005). Kaiso is a bifunctional DNA-binding protein which can recognize DNA sequences containing two methylated CpG dinucleotides (Klose et al., 2005). Only MBD2 so far appears to have an exclusive affinity for mCpG (Allis et al., 2007). DNA methylation and the binding of MBD proteins strongly impact on the modification and structure of chromatin discussed in the next paragraph.

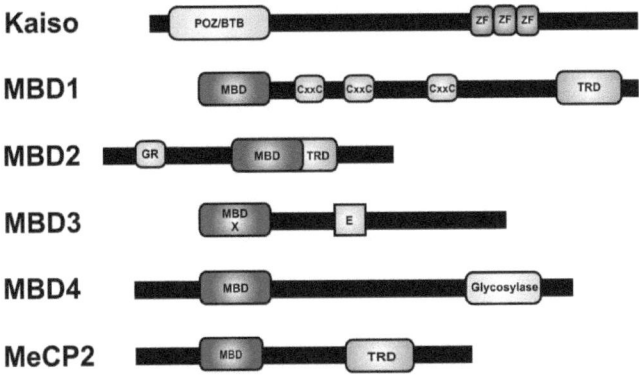

Figure 1-3 Characteristic domains of methyl-CpG binding proteins
Five members of the MBD protein family are aligned at their MBD domains. Other domains are labeled and include transcriptional repression domains (TRD), CXXC domains (zinc fingers some of which are implicated in binding to non-methylated CpG), an E-repeat (E), GR repeats of unknown functions or a T:G mismatch glycosylase domain which is involved in repair of 5-methylcytosine deamination. Kaiso lacks the MBD domain but binds methylated DNA via zinc fingers (ZF) and possesses a POZ/BTB domain to repress transcription (adapted from Klose and Bird, 2006).

1.6.2 Cooperation between DNA methylation and chromatin modifications

In general, the eukaryotic genome is divided into transcriptionally competent euchromatin and transcriptionally incompetent heterochromatin. The nucleosome represents the basic and repeating subunit of chromatin and is composed of a hetero-octamer of histone proteins and 147 bp DNA wrapped around this core 1.7 times in a left-handed helix (Figure 1-4). The histone octamer consists of two H2A-H2B dimers and one $H3_2$-$H4_2$ tetramer and is almost perfectly symmetrical in its tertiary structure (Kornberg, 1974; Kornberg and Lorch, 1999). A single copy of H1 can bind to the 50 bp linker DNA between nucleosomes and plays a significant role in the higher-order packaging of chromatin through stabilizing the chromatin fibre. The position and stability of nucleosomes is a reversible ATP-dependent process. Hence, chromatin is in spite of its strong compaction, a highly dynamic and variable structure. Core histones are highly conserved in eukaryotes and have two subunits: the carboxy-terminal part featuring their common motif, the histone fold, mediates interactions with the DNA. The amino-terminal tails of all eight core histones protrude through the DNA and are exposed on the nucleosomal surface where they are subject to an enormous range of post-translational modifications including acetylation of

Introduction

lysines, methylation of lysines and arginines as well as phosphorylation of serines and threonines (Reid et al., 2009; Turner, 2007). These modifications either allow for improved access for the transcription machinery or the reverse, whereby transcription is prevented in this region due to the conformation of the protein-DNA structure (Bernstein and Allis, 2005; Ducasse and Brown, 2006).

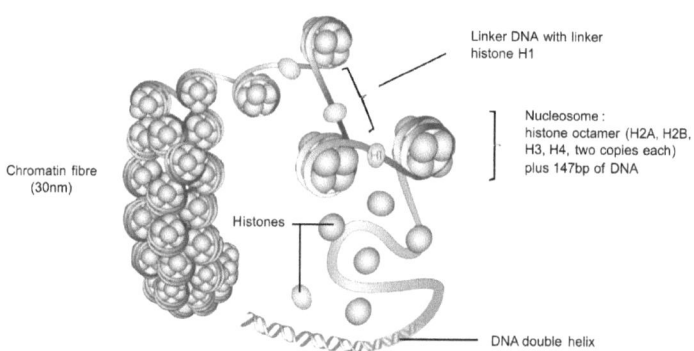

Figure 1-4 DNA compaction into chromatin
In eukaryotes, DNA is packed into chromatin. The basic repeat element of chromatin is the nucleosome, composed of a histone octamer around which 147 bp of DNA are coiled (adapted from Figueiredo et al., 2009).

Recent studies have highlighted the role of DNA methylation in controlling gene expression and have confirmed its links with histone modification and chromatin remodeling (Klose and Bird, 2006). Methyl-binding proteins (MBPs) (see section 1.6.1) act as important "translators" between DNA methylation and histone-modifier proteins since on the one hand they are able to read the epigenetic methyl-CpG code and on the other hand each of the four MBPs has been shown to associate with a different corepressor complex (Lund and van Lohuizen, 2004). For example, MeCP2 interacts with the mSin3a corepressor complex and a histone deacetylase (HDAC). Besides, it is also able to recruit DNMT1 to promoters (Ballestar and Wolffe, 2001; Jones et al., 1998; Kimura and Shiota, 2003; Nan et al., 1998). Of particular interest is MBD1, which can associate with the histone H3 lysine 9 (H3K9) methyltransferase SETDB1 only during DNA replication (Sarraf and Stancheva, 2004). MBD2 is the DNA-binding component of MeCP1, which additionally includes the NuRD (nucleosome remodeling and histone deacetylation) (or Mi-2) corepressor complex (Wade et al., 1999). NuRD comprises MTA2 (metastasis-associated

Introduction

protein), MBD3, the histone deacetylases HDAC1 and HDAC2, a large chromatin-remodeling protein (Mi-2) and RbAp46/48, a component of several chromatin-related processes (Feng and Zhang, 2001; Loyola and Almouzni, 2004) (Allis et al., 2007).
Histone modifications and chromatin remodeling can block transcription factors whereby a transcriptionally inactive chromatin environment is established (Esteller, 2007b). The most important processes concerning histone modifications will be focused in more detail in the following sections. Additionally, an example of the cooperation between DNA methylation and chromatin modification is summarized in Figure 1-5.

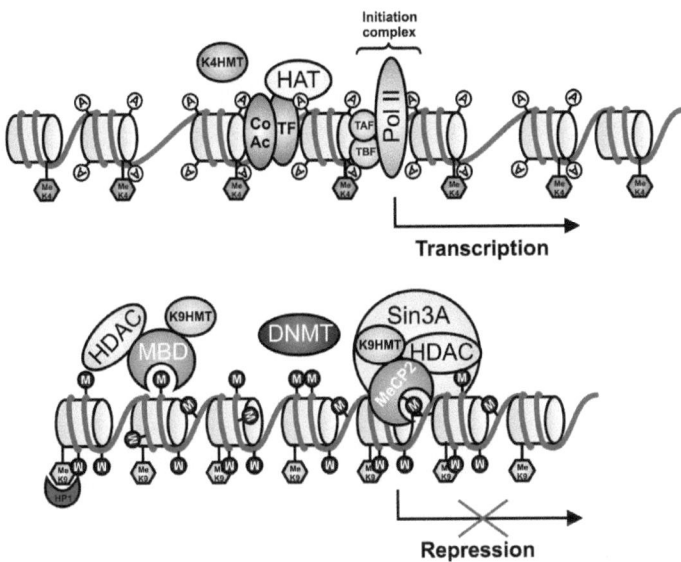

Figure 1-5 DNA methylation, chromatin structure and recruitment of multiple repressors in a hypermethylated CpG island
The open chromatin structure of a transcriptionally active gene with loosely spaced nucleosomes (cylinders) marked by DNA demethylation, histone acetylation (A) and histone H3K4 methylation is shown at the top. The transcriptionally silenced state with more tightly packed nucleosomes is shown at the bottom. In formation of heterochromatic structures MBDs, HDACs, DNMTs and H1 are involved. MeCP2 is believed to recruit the Sin3A HDAC complex and HMT activity to the methylated site. Histone acetylation is indicated by circles labeled with an "A", H3K4 and H3K9 methylation is indicated by hexagons and methylated CpG dinucleotides are indicated by circles marked with an "M". Proteins involved in transcriptional activation: Pol II=DNA polymerase II; TF=transcription factor; CoA=coactivator, HAT=histone acetyltransferase; TBF=TATA-binding factor; TAF=TBP-associated factor; Histone H3 lysine 4 methyltransferase is abbreviated as K4 HMT. Proteins involved in transcriptional silencing: DNMT=DNA methyltransferase; MBD and MeCP2=methyl binding domains; HP1=heterochromatin protein 1; Histone H3 lysine 9 methyltransferase is abbreviated as K9 HMT (adapted from Allis et al., 2007; Laird, 2005).

1.6.3 The histone code

The so-called histone code as part of the epigenetic code, comprises multiple histone modifications which act sequentially or in a combination either on one or on multiple histone tails and thereby specify unique downstream functions (de, X et al., 2005; Strahl and Allis, 2000). Recent discoveries showed that the functional epigenetic landscape is much more complex than previously thought which led to a refining of the histone code hypothesis. One aspect is that specific histone marks can have either repressive or activating consequences depending on the influence of adjacent modifications (de, X et al., 2005). For example, methylation of histone 3 lysine 9 (H3K9me) can initiate gene silencing but, in the context of methylated H3K4 and H4K20 it helps maintaining active marks (de, X et al., 2005). Likewise, H3K36 has a positive effect on transcription when it is found on the coding region and a negative effect when it is located inside the promoter sequence. Furthermore the data revealed that modifications on the same or different histones may be interdependent (de, X et al., 2005). That means that modification in one residue can determine that of another one either in *cis* or also in *trans* (de, X et al., 2005). An example for *cis* effects is represented by the activating mark H3K4me, which has two consequences: disrupting the binding of the repressive NuRD complex as well as blocking the methylation of H3K9. The best studied example for a *trans* effect is the ubiquitination of H2B being required for methylation of H3K4me3 (de, X et al., 2005; Kouzarides, 2007).

Consequently, a specific histone mark alone does not describe a specific transcriptional state (active or passive), which turns transcription on or off, respectively. Actually, the marks have to be read in the context and in combination within the landscape of all the other marks decorating the chromatin platform and can thus represent a mechanism for differential regulation of chromatin activity in several distinct biological settings (Berger, 2007; Strahl and Allis, 2000; Weissmann and Lyko, 2003).

Within the last few years there has been considerable progress in the development of high-throughput methods for analyzing histone modifications. Systematic and extensive studies of chromatin modifications performed either by mass spectrometry, ChIP-on-chip experiments or sequencing methods revealed a complex landscape including clusters of modified histones at transcription start sites, distal regulatory elements and conserved sequences, and broad domains at gene clusters and developmental loci (Bernstein et al., 2007). Altogether at least eight distinct types of modifications on over 60 different histone residues were identified (Kouzarides, 2007). The most prominent ones are illustrated in Figure 1-6.

Introduction

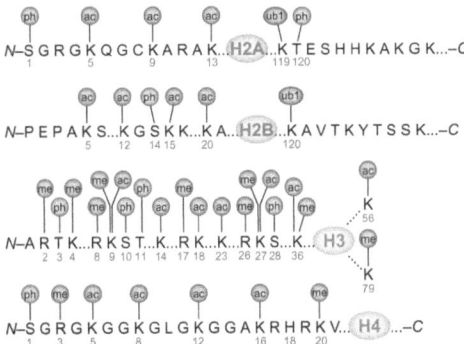

Figure 1-6 Post-translational histone modifications
The modifications include acetylation (ac), methylation (me) and phosphorylation (ph) on lysine (K), arginine (R), serine (S) and threonine (T) residues. Histone modifications occur mainly on the N-terminal tails of histones H2A, H2B, H3 and H4 (adapted from Bhaumik et al., 2007).

While the combination of all different histone modifications is an important aspect of epigenetic gene regulation, the remainder of this chapter will focus on histone acetylation and methylation, and how they relate to DNA methylation and gene expression.

1.6.3.1 Histone acetylation

Histone acetylation promotes transcriptionally active chromatin states by neutralizing the basic charge of the lysine residues, which weakens the interaction between the DNA and histone proteins, as well as between neighboring nucleosomes (Kouzarides, 2007). Acetylation occurs by the action of histone acetyltransferases (HAT). HATs are divided into three main families, GNAT, MYST and CBP/p300 that do not show much preference for a specific lysine residue generally (Kouzarides, 2007). Most of the acetylation sites are located on the histone tails, with the exception of lysine K56 located within the core domain of histone H3. K56 is facing towards the major groove of the DNA and can therefore strongly affect histone-DNA interactions when acetylated (Kouzarides, 2007).
The antagonists to histone acetylases represent the histone deacetylases (HDAC) which remove the acetylation marks from the lysine residues. Thereby the positive charge of the histones is restored and therefore interacts with the negative charges on the DNA-phosphate backbone resulting in a more condensed chromatin structure. There are three distinct families of HDACs described: class I and class II HDACs and class III NAD-dependent enzymes of the Sir family (Kouzarides, 2007). HDACs have been found to

be associated with transcriptional repressor complexes (see section 1.6.2). In addition, HDACs are able to interact directly with transcription factors like YY1 or the nuclear corepressor NCoR, as well as with other HDACs (Dobrovic and Kristensen, 2009). Therefore, in addition to inducing a closed chromatin structure, HDACs are co-recruited with other proteins which block transcription.

1.6.3.2 Histone methylation

While histone acetylation to date has only been found associated with gene activation, histone methylation may have either activating or repressive effects on transcription depending on the specific residue modified and the context of other modifications. Additional complexity comes from the fact that lysine but also arginine residues can be methylated to different extents by histone methyltransferases (HMTs): lysine can be mono-, di- and trimethylated and arginine can be mono- or dimethylated, both, symmetrically and unsymmetrically (Kouzarides, 2007). All three states of H3K4 methylation are characteristic features of gene expression. Trimethylation of histone H3 lysine 36 (H3K36me3) and monomethylation of H3 lysine 27 (H3K27me1), H3 lysine 9 (H3K9me1), H3 lysine 20 (H3K20me1), H3 lysine 79 (H3K79me1) and H2B lysine 5 (H2BK5me1) are also associated with transcribed chromatin. In contrast, trimethylation of H3 lysine 9 (H3K9me3), H3 lysine 27 (H3K27me3) and H3 lysine 79 (H3K79me3) is generally linked to repression (Barski et al., 2007; Bernstein et al., 2007).

It was long believed that histone methylation was irreversible and thus the only stable histone modification. However, the recent discovery of histone demethylases has shown that histone methylation is as dynamic as the other histone modifications. Currently, there are two known types of histone demethylase domains: the LSD1 domain and the JmjC domain. Contrary to histone acetyltransferases, the histone methyltransferases as well as the histone demethylases show a high degree of substrate specifity, which is possibly the reason why methylation is currently the best characterized modification (Kouzarides, 2007).

1.6.3.3 Recognition of chromatin modifications and the translation of the histone code

The functional consequences of histone modifications can be either direct, causing structural changes to chromatin, or indirect, acting through the recruitment of effector proteins (Berger, 2007). There are two main classes of proteins that can interact with specific chromatin modifications and bind via specific domains (Kouzarides, 2007). While methylation is recognized by so-called chromodomains, acetylation is recognized by bromodomains (Kouzarides, 2007).

Bromodomains are cysteine-rich motifs which facilitate protein-protein interactions and were found to be widely distributed among the different enzymes that acetylate (e.g.GCN5/PCAF, PCAF (CBP/300), $TAF_{II}250$, TAF1I), methylate (e.g. MLL, a member of the TRX proteins) or remodel (SWI/SNF complex) chromatin (Daniel et al., 2005; de, X et al., 2005; Kouzarides, 2007; Taverna et al., 2007). Remodeling factors may promote transcription by moving away blocking nucleosomes from transcription factor binding sites, as has been described for the Mi-2/NuRD and SWI/SNF complexes discovered in yeast (Hassan et al., 2002; Jacobson et al., 2000).

The chromodomain was first identified as a common domain in HP1 (chromodomain-containing heterochromatin protein 1) and the Polycomb protein of *Drosophila* (de, X et al., 2005). Later, chromodomains have also been detected in many other chromatin regulators like in ATP-dependent chromatin-remodeling enzymes (BPTF, CHD1, RAD54, Mi-2), HATs (ING2, MORF4L1) and HMTs (SUV39H1 and H2). Recently, it was shown that the HP1 chromodomain can recognize methylation of H3K9 (Bernstein et al., 2007) which induces transcriptional repression and heterochromatinization (Bartova et al., 2008). HP1 is associated with deacetylase and methyltransferase activity. Another example are the Polycomb (PcG) and trithorax (TrxG) group proteins that function as antagonistic chromatin-modifying complexes. They operate through binding to *cis*-acting PcG responsive elements (PREs) and form the molecular basis of the cellular memory. TrxG is required for the active state, whereas PcG proteins mediate the repressed state of gene expression. PcG proteins play pivotal roles in development and in the epigenetic silencing of lineage-specific gene repression. They are required for embryonic stem (ES) cell pluripotency and are markedly downregulated upon differentiation. PcG proteins are divided into two families based on distinct Polycomb repressor complexes, namely PRC1 and PRC2. PRC proteins are recruited to their response elements. PCR2 modifies the chromatin by catalyzing H3K27 and H3K9 methylation, while PCR1 complexes create stably repressed chromatin structure through recognition of H3K27me3 via its chromodomain protein PC, in analogy to the formation of constitutive heterochromatin (Muller et al., 2002; Peters and Schubeler, 2005; Ringrose and Paro, 2007).

However, effector proteins and complexes often contain multiple modification binding domains, with the potential to bind adjacent marks either within one histone or among multiple nucleosomes. HP1, for example, may function as a dimer that binds two methylated sites (Rice and Allis, 2001). Figure 1-7 illustrates the function of conserved motifs with certain chromatin-modifying proteins (Rice and Allis, 2001).

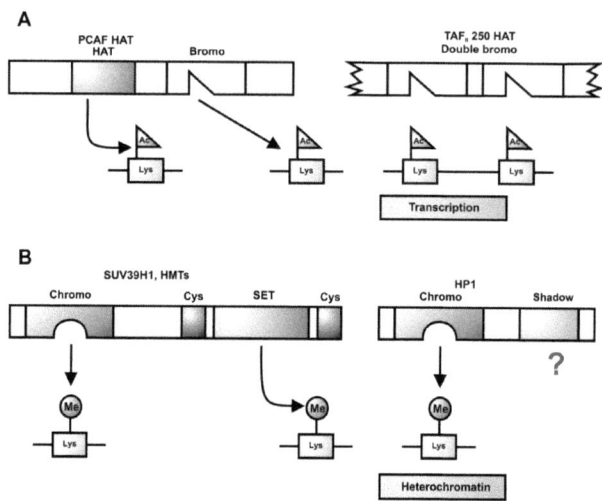

Figure 1-7 Schematic representation of the function of bromo- and chromodomains
(A) PCAF and TAF$_{II}$250 contain a HAT catalytic domain that may acetylate lysine residues on the histone tails (not shown for TAF$_{II}$250). Additionally, each protein contains a bromodomain or double bromodomain, respectively, that binds to the acetylated lysines on the histone tails to promote transcription. (B) SUV39H1 contains a catalytic SET domain flanked by two cysteine-rich domains (Cys) which are necessary for methyltransferase activity. The chromodomain of HP1 binds to specific methylated histone tails such as H3K9 and induces the assembly of heterochromatin. The exact functions of the HMT chromodomain and HP1 chromo shadow domain are not known (adapted from Rice and Allis, 2001).

In summary, chromatin provides a platform that becomes regulated by structural marks which can be read by nuclear factors. In order to act as marks which can influence the chromatin structure and thereby the transcriptional state of a gene, modifications have to be directed to the specific loci. There are several ways of targeting modifying enzymes to their sites of action (Imhof, 2006). One possibility is the targeting through interaction with specific transcription factors. Moreover, histone modifying enzymes have also been shown to interact with RNA polymerases or the replication clamp proliferating cell nuclear antigen (PCNA) (Imhof, 2006). Another targeting mechanism is the recruitment of histone deacetylases by methyl binding proteins (MBPs) to methylated cytosines. Recently, another mechanism of targeting histone modifying activities has been proposed involving the transcription of non-coding RNAs (Imhof, 2006). The non-coding *Xist* RNA, for example, coats the entire inactive X chromosome, causing chromosome-wide gene silencing. This process is accompanied by the deposition of histone modifications like H3K27me3 and H4K20me1 (Bartova et al., 2008).

The ability of the histone code to dictate the chromatin environment allows not only the regulation of transcriptional activity but also the regulation of other nuclear processes such as replication, DNA repair, and chromosome condensation (Kouzarides, 2007).

1.6.4 Non-coding RNA

Recent studies have demonstrated that non-coding RNAs (ncRNAs) such as miRNAs act as diverse players in gene regulation, especially in the epigenetic control of chromatin. ncRNAs are able to direct methylation of CpGs as well as histone modifications that are correlated to long-term gene silencing (Costa, 2008). In a yeast model, Moazed et al. demonstrated that components of the RNAi (RNA interference) participate directly in heterochromatin formation (Moazed et al., 2005; Moazed, 2007). Therefore, it was proposed that the nascent RNA transcripts from centromeric repeats may act as a platform for heterochromatin assembly. Liu et al. could show by knockout experiments with *Tetrahymena* that H3K27me1 (a mark for repressive heterochromatin) is dependent on the RNAi machinery (Liu et al., 2007). This provides an indication that ncRNAs may mediate the heritability of histone modifications and heterochromatin formation (Flanagan, 2007).

One of the best studied examples of ncRNAs involvement is the dosage compensation through silencing of the second X chromosome by the ncRNA Xist as described above (Bernstein and Allis, 2005).

Although, the knowledge about the influence of non-coding RNA on transcriptional changes is far from being complete, those molecules are considered to be important epigenetic regulators.

1.7 Epigenetic alterations during tumorigenesis

Cancerogenesis constitutes a multistep process in which defects in various tumor genes accumulate (Plass, 2002). The initiation and progression of cancer is due to genetic changes such as point mutations, missense or frameshift mutations, deletions and translocations, but also to epigenetic changes (Herman and Baylin, 2003). Epigenetic tumor-specific alterations comprise most importantly DNA methylation as well as histone modifications which can influence gene regulation of oncogenes or tumor suppressor

Introduction

genes and contribute to uncontrolled cell growth (Costello and Plass, 2001; Plass, 2002). DNA methylation changes in cancer cells include both loss of methylation in CpG depleted regions where most CpGs should be methylated (hypomethylation) or gains of methylated CpGs in CpG islands (hypermethylation) (Herman and Baylin, 2003; Plass, 2002). Figure 1-8 summarizes the different mechanisms through which DNA methylation can promote oncogenesis.

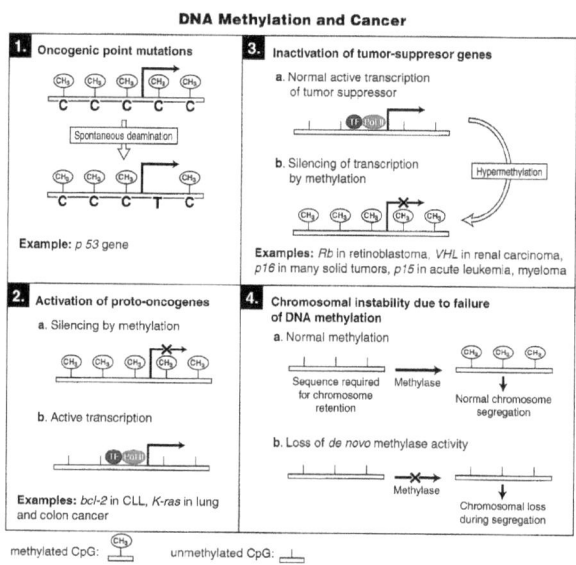

Figure 1-8 Models for the different mechanisms through which cytosine methylation can promote oncogenesis
(1) A consequence of hydrolytic deamination of 5mC are cytosine to thymine transitions. Those point mutations within promoters of tumor suppressor genes (if both alleles are affected) may contribute to tumorigenesis. (2) Specific oncogenes have been observed to be hypomethylated and maybe therefore activated in human tumors. (3) Tumor suppressor genes can be inactivated through promoter hypermethylation. (4) Loss of methylation may contribute to chromosome instability which possibly leads to gene deletions during mitotic recombination (adapted from Singal and Ginder, 1999).

1.7.1 Global hypomethylation

A major hallmark in cancer is the occurrence of genome-wide hypomethylation (Plass, 2002; Singal and Ginder, 1999). The extent of global hypomethylation is correlated to the tumor's malignancy grade. Therefore, decreased levels of overall genomic methylation may serve as biological marker with prognostic value (Costello and Plass, 2001). The majority of hypomethylation events occur in repetitive elements localized in satellite sequences or centromeric regions (Plass, 2002). Furthermore, hypomethylation contributes to the activation of latent retrotransposons and to the potentially harmful expression of inserted viral genes, imprinted genes and genes on the inactive X chromosome (Costello and Plass, 2001; Herman and Baylin, 2003). In addition, the global loss of DNA methylation affects the functional stability of chromosomes in cancer. Pericentromeric regions of the chromosomes depend on high levels of cytosine methylation for stability and for proper replication of the DNA (Herman and Baylin, 2003). Aside from the genome-wide hypomethylation, the demethylation and consequently the activation of specific proto-oncogenes have been reported (Singal and Ginder, 1999). Proto-oncogenes are normal genes that can become oncogenes due to mutations or increased expression. They usually code for proteins that are either involved in regulation of cell growth and differentiation or in signal transduction. As oncogenic mutations are dominant, only one mutated allele is necessary to confer the cancerous behavior. Some well described examples of proto-oncogenes include *Ras, bcl-2 or c-myc* (Costello and Plass, 2001).

1.7.2 Regional hypermethylation

CpG islands located in the promoter region of certain genes fulfill gene regulatory functions and are generally unmethylated in healthy cells. Exceptions are imprinted genes and genes on the inactive X chromosome. There is also evidence that selected genes show progressive age-related increases in promoter methylation (Issa, 2003). Particularly, in cancer cells, prevalently CpG islands (CGIs) of tumor suppressor genes can be hypermethylated despite genome-wide hypomethylation (Hirst and Marra, 2009). Aberrant *de novo* CGI promoter methylation leads to gene silencing affecting genes involved in cell cycle, DNA repair, metabolism, cell adherence, apoptosis, premature aging and miRNA expression (Esteller, 2007b; Hirst and Marra, 2009). The Retinoblastoma (*Rb*) gene was the first tumor suppressor gene targeted by CGI hypermethylation. Another tumor suppressor gene frequently found hypermethylated in cancer is *p16*, an important cell cycle regulatory protein. P16 (also known als INK4a or CDKN2A) is responsible for blocking cell cycle progression at the G1/S boundary. Loss of p16 function through methylation may lead to cancer progression by allowing deregulated cellular proliferation (Singal and Ginder, 1999). The number of genes that are known to be affected by epigenetic inactivation exceeds the number of cancer-related genes inactivated by mutation (Herman and Baylin, 2003). Mutations in tumor suppressor genes are mostly recessive which explains why the complete disruption of a tumor suppressor gene's function requires inactivation of both alleles for malignant transformation of a cell (Costello and Plass, 2001; Herman and Baylin, 2003).

So far, the mechanism for tumor-specific hypermethylation is not yet fully explored. The profiles of CpG island hypermethylation of tumor suppressor genes vary according to the tumor type (Esteller, 2005; Esteller, 2007b). Different mechanisms for aberrant *de novo* methylation in cancer may have evolved. One possibility is that methylation changes arise in a random fashion and may lead to progressive proliferation through a selective advantage (Jones and Baylin, 2007). The second possibility constitutes a dysregulation of histone and DNA modifying enzymes such as DNMTs. The third possibility for abnormal *de novo* methylation in tumor cells includes the absence of "protective" transcription factors or a loss of chromatin boundaries leading to the spreading of DNA methylation into affected CpG islands (Turker, 2002). The fourth possibility comprises the targeted recruitment of DNMTs to methylation targets by interaction with transcription factors such as the Ets family transcription factor PU.1 (Metivier et al., 2008; Suzuki et al., 2006) or the SET-containing histone methyltransferases like G9a (Feldman et al., 2006) or EZH2. The histone methyltransferase EZH2 specifically methylates both H3K9 and H3K27. EZH2 is associated with the Polycomb group (PcG) complexes (see section 1.6.3.3 and 1.7.4), and is recruited to PcG-specific sites through SUZ12 (Vire et al., 2006). EZH2 has emerged as a key histone methyltransferase involved in methylation of H3K27 within promoters of developmental genes that become methylated and therefore reversibly repressed to maintain pluripotency in ES cells. It was shown that stem cell PcG targets are more likely affected by cancer-specific *de novo* methylation than non-PcG targets during malignant transformation (Schlesinger et al., 2007). This finding supports the hypothesis that reversible gene expression in a stem cell is replaced by permanent silencing in a cancer cell with perpetual state of self-renewal (Widschwendter et al., 2007). This is possibly due to the upregulation of EZH2 observed in many tumors. Recent studies have described a direct connection between EZH2 and the DNMTs (Vire et al., 2006), thus providing a link between histone methylation and DNA methylation during cancer development.

1.7.3 Differential DNA methylation patterns in AML and colorectal cancer

In addition to the large amount of well defined genetic aberrations as described previously, DNA methylation changes at crucial genes are able to contribute to the multistep transformation process of normal to cancerous cells (Farrell, 2005; Galm et al., 2006; Pfeifer and Besaratinia, 2009; Plass et al., 2008). Although some genes are affected in multiple tumor types, such as the cell cycle regulator p16, methylation profiling studies have shown that each tumor type has a characteristic methylation pattern indicating the involvement of cell type- or lineage-specific transcription factors. Interestingly, the methylation profile in hematopoietic malignancies differs from solid tumors regarding the affected genes. Table 1-2 gives a summary of those genes that have frequently reported to be hypermethylated in AML and in colorectal carcinoma, respectively.

There is growing evidence for the diagnostic and prognostic potential of methylation changes in different tumor types. Methylation profiles may act as potential new biomarkers of risk prediction in tumor patients, complementing standard immunophenotyping, cytogenetic and molecular analyses (Galm et al., 2006; Plass et al., 2008).

Table 1-2 Genes frequently methylated in acute myeloid leukemia (AML) and colorectal carcinoma
(adapted from Galm et al., 2006; Toyota et al., 2001; Wong et al., 2007)

AML		Colorectal carcinoma	
Gene	Function	Gene	Function
p15	Cyclin-dependent kinase inhibitor	APC	Signal transduction, β-catenin regulation
E-cadheri	Cell adhesion	CDH13	Cell signaling
SOCS-1	Cell signaling	CDKN2A	Cell cycle regulation
p73	p53 similar protein	CHFR	Mitotic stress checkpoint
DAPK1	Programmed cell death induced by IFNγ	HIC1	Regulation of DNA damage responses
HIC1	Transcriptional regulation	HPP1	TGF-β antagonist
RARβ2	Retinoic acid receptor-β2	MGMT	Repair of DNA guanosine methyl adduct
CRBP1	Carrier protein involved in retinol transport	MLH1	Mismatch repair
MYOD1	Myogenic differentiation	RASSF1,	DNA repair, cell cycle regulation
SDC4	Receptor in intracellular signaling	TIMP3	Matrix remodeling, tissue invasion

1.7.4 Differential histone modifications in tumors

Global changes in the pattern of CpG methylation eventually lead to genome-wide changes in histone modification patterns in human tumor cells (Fraga et al., 2005). These alterations of histone marks occur early in tumorigenesis and accumulate during its process (Esteller, 2007a).

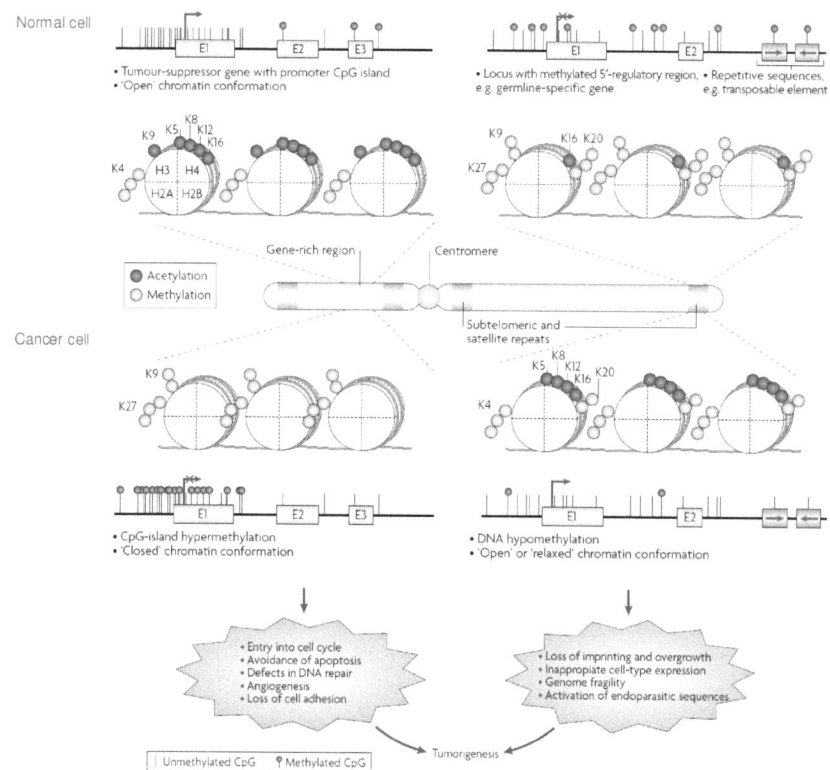

Figure 1-9 Histone modification maps and DNA methylation patterns for a typical chromosome in normal and cancer cells
Nucleosomes (grey cylinders) are shown in the context of chromosomal location and transcriptional activity. Histone acetylation and methylation (di- and tri-) are indicated. In normal cells, genomic regions that include the promoters of tumor suppressor genes are enriched for histone modification marks associated with active transcription, such as acetylation of H3 and H4 lysine residues and H3K4me3. In the same cells, DNA repeats and other heterochromatic regions are characterized by H3K27me3, H3K9me2 and H4K20me3 which are repressive marks. In cancer cells, both, the active histone marks on tumor suppressor genes and repressive marks at DNA repeat regions are lost. Above and below the nucleosomal arrays, respectively, the respective DNA methylation pattern is represented. CpG islands of tumor suppressor genes become hypermethylated in cancer cells which leads to transcriptional inactivation of these genes. At the same time, the genome of a cancer cell undergoes global hypomethylation at repetitive sequences and some tissue-specific and imprinted genes. This hypomethylation might contribute to tumorigenesis, causing chromosomal instability or changes such as loss of imprinting. E, exon (adapted from Esteller, 2007a).

Introduction

Promoter hypermethylation in cancer cells is generally linked to changes in the histone modification pattern including hypoacetylation at histones H3 and H4 (Esteller, 2006). Furthermore, in some cases H3K9 methylation has been shown to occur in hypermethylated DNA sequences. Usually, aberrant DNA hypermethylation is also accompagnied by loss of H3K4 trimethylation (Esteller, 2006).

In cancer cells, altered patterns of histone modifications occur not only in promoter regions of tumor suppressor genes but also within constitutive heterochromatin. Studies of genome-wide posttranslational modifications of histone H4 have shown that cancer cells exhibit decreased levels of H4K16 acetylation and H4K20 trimethylation compared to normal tissues (Esteller, 2007a; Fraga et al., 2005; Fraga and Esteller, 2005). Those alterations occur mainly within the context of repetitive DNA sequences that become hypomethylated in tumor cells. It was thus suggested that the global loss of monoacetylation and trimethylation of H4 is a common hallmark of cancer cells (Fraga et al., 2005; Fraga and Esteller, 2005). A summary of altered histone modifications and DNA methylation patterns is shown in Figure 1-9.

Recent analyses have shown that a number of different histone modifying enzymes are altered in various types of cancer contributing to changes in histone patterns. The observed loss of H4K20me3 could be due to altered expression levels of histone methyltransferases (like EZH2, SUV39H or SUV4-20H) (Esteller, 2007a). The proper balance between acetylation and deacetylation of H4K16 is mediated by specific HATs. Inactivating mutations in the HAT p300, for example, have been described in many tumor forms. Recently, HDAC2 has been identified as another component of the epigenetic machinery that is mutated and therefore inactivated in human cancer (Esteller, 2007a).

One of the most interesting phenomena currently emerging in the field of histone modifications and cancer is the discovery of bivalent domains on histone tails. Bivalent domains comprise both, transcriptionally repressive and active marks at the same time, which facilitates a switch between transcriptional activation and repression. Those domains were first described in ES cells (Bernstein et al., 2006). Upon differentiation, tissue-specific genes which become highly expressed display only the activating H3K4 methylation, while genes which become silenced display only the repressive H3K27 methylation (Bernstein et al., 2006; Pietersen and van Lohuizen, 2008). Bivalent domains serve to silence developmental genes in ES cells, while preserving the potential for

Introduction

activating them during differentiation (Bernstein et al., 2006; Pietersen and van Lohuizen, 2008) (see sections 1.6.3.3 and 1.7.2).
It is believed that genes with bivalent marks are especially sensitive to silencing by DNA methylation during tumorigenesis (Ohm et al., 2007). Notably, embryonal carcinoma cells (EC) also display increased levels of other repressive marks (H3K9 di-and tri-methylation) on the histones associated with the respective genes, implicating that these bivalent histone marks represent a transition state between genes in ES cells and fully silenced genes in adult cancer cells (Ohm et al., 2007).

1.7.5 Therapeutic strategies targeting epigenetic aberrations

In contrast to genetic alterations, epigenetic changes which contribute to tumorigenesis are potentially reversible. This offers the possibility of novel therapies for cancer treatment, particularly with regard to acute myeloid leukemia (AML) and myelodysplastic syndrome (MDS) (Galm et al., 2006; Herman and Baylin, 2003). Many studies in multiple tumors demonstrated an increased activity of DNMTs as well HDACs. Therefore, the development of strategies will be required which inhibit and antagonize (reverse) the activity of those enzymes in order to prevent and treat neoplastic diseases (Herman and Baylin, 2003; Singal and Ginder, 1999).
Currently, there are two demethylating agents approved for clinical use by the United States FDA: 5-azacytidine (Vidaza, AZA; Pharmion Corp) and 5-aza-2´-deoxycytidine (decitabine, Dacogen; MGI Pharma/SuperGen) (Herman and Baylin, 2003; Plass et al., 2008). These agents are cytosine analogues and exert their hypomethylating activity by competing with the endogenous pool of deoxynucleosides for incorporation into newly synthesized DNA during replication. Once incorporated into DNA, azanucleosides covalently bind and trap the DNMTs resulting in inactivation. This may lead to loss of promoter hypermethylation and re-expression of silenced tumor suppressor genes (Plass et al., 2008).
Previous studies demonstrated the dominance of the DNA methylation over histone deacetylation in the process of gene silencing. Thus, treatment of cells with one of the HDAC inhibitors such as Trichostatin A, valproic acid, the hydroxamid acid or SAHA (Vorinostat), alone, rarely results in reactivation of cancer genes (Plass et al., 2008). However, these agents exert additive or synergistic effects if some demethylation is first achieved by low doses of 5-aza-2´-deoxycytidine (Galm et al., 2006; Herman and Baylin,

2003). *In vivo,* the sequential treatment with low doses of demethylating agents followed by HDAC inhibitors is the basis for new therapeutic strategies.

Recent studies have shown that epigenetic mutations may be more harmful than genetic mutations. Thus, patients might benefit from the new emerging anti-cancer treatments which target the epigenome. Additionally, because many of the hypermethylation events occur very early in tumor development, inhibiting or reversing these changes could be of high potential for cancer prevention in the future.

2 Research objectives

Besides genetic alterations, epigenetic changes are now recognized as an additional mechanism contributing to tumorigenesis (Plass, 2002). Because DNA methylation is stable in genomic DNA preparations, it is the most suitable of all known epigenetic modifications for diagnostic applications and may provide useful molecular markers to complement clinical diagnostics and prognostics. To date, most of the studies are based on single gene approaches to identify candidate genes. However, for the systematical identification of relevant epigenetic biomarkers global analyses of DNA methylation are of major clinical interest.

The main focus of this thesis was to establish and adapt the previously developed methyl-CpG immunoprecipitation (MCIp) technique for comparative methylation analyses. Based on this approach, genome-wide methylation profiles of tumor cell lines and tumor patients should be generated, to detect potential marker genes which are hypermethylated in tumors and could be associated with cancer development and therefore have diagnostic or therapeutic relevance. Analyses should be performed with two specific tumor types, namely acute myeloid leukemia (AML) and colon cancer carcinoma.

The molecular mechanisms controlling the methylation status of CpG islands in normal and malignant cells are poorly understood. Therefore, to get insights into this process, factors should be identified which are mainly responsible on the one hand for maintaining the unmethylated state of CpG islands in health and disease and on the other hand for *de novo* methylation in cancer. Computational methods should be applied to identify candidate sequence motifs associated with unmethylated and methylated CpG islands, respectively.

3 Material and equipment

3.1 Equipment

8-Channel PipettorImpact2 Equalizer 384	Thermo Fisher Scientific, Hudson, US
Autoclave	Technomara, Fernwald, Germany
Bioanalyzer 2100	Agilent Technologies, Böblingen, Germany
BioPhotometer	Eppendorf, Hamburg, Germany
Centrifuges	Heraeus, Hanau; Eppendorf, Hamburg
Densitometer	Molecular Dynamics, Krefeld, Germany
Electrophoresis equipment	Biometra, Göttingen; BioRad, Munich, Germany
Fast-Blot machine	Agfa, Köln
FACS Calibur	BD, Heidelberg, Germany
Heat sealer (Fermant 400)	Josten & Kettenbaum, Pensberg, Germany
Heat sealer	Eppendorf, Hamburg, Germany
Heatblock	Stuart Scientific, Staffordshire, UK
Incubators	Heraeus, Hanau, Germany
J6M-E centrifuge	Beckmann, Munich, Germany
Laminar air flow cabinet	Heraeus, Hanau, Germany
Lightcycler	Roche, Mannheim
Luminometer (Sirius)	Berthold Detect. Systems, Pforzheim, Germany
MassARRAY Compact System	Sequenom, Hamburg, Germany
MassARRAY MATRIX Liquid Handler	Sequenom, Hamburg, Germany
MassARRAY Phusio chip module	Sequenom, Hamburg, Germany
Megafuge 3,0 R	Heraeus, Osterode, Germany
Microarray hybridization chambers SureHyb	Agilent Technologies, Böblingen, Germany
Microarray scanner; 5 micron resolution	Agilent Technologies, Böblingen, Germany
Microarray slide holder	Agilent Technologies, Böblingen, Germany
Microscopes	Zeiss, Jena, Germany
Multifuge 3S-R	Heraeus, Osterode, Germany
Multipipettor Multipette plus	Eppendorf, Hamburg, Germany
NanoDrop	PeqLab, Erlangen, Germany
PCR-Thermocycler PTC-200	MJ-Research/Biometra, Oldendorf, Germany
PCR-Thermocycler Veriti 384 well	Applied Biosystems, Foster City, USA
pH-Meter	Knick, Berlin, Germany
Picofuge	Heraeus, Osterode, Germany
Power supplies	Biometra, Göttingen; Germany

Material and equipment

Realplex Mastercycler epGradient S	Eppendorf, Hamburg, Germany
Sigma 2 – Sartorius	Sartorius, Göttingen, Germany
Sonifier 250	Branson, Danbury, USA
Sorvall RC 6 plus	Thermo Fisher Scientific, Hudson, USA
Speed Vac	Christ, Osterode, Germany
Thermomixer	Eppendorf, Hamburg, Germany
Typhoon™	Amersham Biosciences, Germany
Ultracentrifuge Optima L-70	Beckman, Munich, Germany
Waterbath	Julabo, Seelstadt, Germany
Water purification system	Millipore, Eschborn, Germany

3.2 Consumables

384-well PCR plates	Thermo Fisher Scientific, Hudson, USA
96 well optical bottom plates (black)	Nunc Brand Products, Roskilde, Denmark
8-channel pipettor tips Impact 384	Thermo Fisher Scientific, Hudson, USA
Adhesive PCR sealing film	Thermo Fisher Scientific, Hudson, USA
Cell culture flasks and pipettes	Costar, Cambridge, USA
CLEAN resin	Sequenom, Hamburg, Germany
Cryo tubes	Nunc, Wiesbaden, Germany
Filter tubes: Millipore Ultrafree-MC	Millipore, Eschborn, Germany
Heat sealing film	Eppendorf, Hamburg, Germany
Luminometer vials	Falcon, Heidelberg, Germany
MATRIX Liquid Handler D.A.R.Ts tips	Thermo Fisher Scientific, Hudson, USA
Micro test tubes (0.5, 1.5, 2 ml)	Eppendorf, Hamburg, Germany
Microarray gasket slides	Agilent Technologies, Santa Clara, USA
Multiwell cell culture plates and tubes	Falcon, Heidelberg, Germany
nProteinA Sepharose 4 FastFlow	GE Healthcare, Munich, Germany
Nylon Transfer membrane	MSI, Westboro, USA
PCR plate Twin.tec 96 well	Eppendorf, Hamburg, Germany
rProteinA Sepharose 4 FastFlow	GE Healthcare, Munich, Germany
Sepharose Cl-4 beads	Sigma-Aldrich, Munich, Germany
SpectroCHIP bead array	Sequenom, Hamburg, Germany
Sterile combitips for Eppendorf multipette	Eppendorf, Hamburg, Germany
Sterile micropore filters	Millipore, Eschborn, Germany
Sterile plastic pipettes	Costar, Cambridge, USA
Syringes and needles	Becton Dickinson, Heidelberg, Germany

3.3 Chemicals

All reagents used were purchased from Sigma-Aldrich (Taufkirchen, Germany) or Merck (Darmstadt, Germany) unless otherwise noted. Oligonucleotides for real-time PCR were synthesized and high-pressure liquid chromatography purified by Metabion (Planegg-Martinsried, Germany). Oligonucleotides adapted to methylation analysis with the MassARRAY system (see section 4.4.6) were purchased from Sigma-Aldrich (Taufkirchen, Germany).

3.4 Enzymes and kits

aCGH Hybridization Kit	Agilent Technologies, Waldbronn, Germany
Alkaline phosphatase	Roche, Mannheim, Germany
Aprotinin	Roche, Mannheim, Germany
BioPrime Purification Module	Invitrogen, Karlsruhe, Germany
BioPrime Total Genomic Labelling System	Invitrogen, Karlsruhe, Germany
Blood & Cell Culture DNA Midi Kit	Qiagen, Hilden, Germany
BSA	Sigma, Deisenhofen, Germany
Blood and Tissue Culture Kit	Qiagen, Hilden, Germany
DNA Ladder 1 kb plus	Invitrogen, Karlsruhe, Germany
DNA molecular weight standard	Invitrogen, Karlsruhe, Germany
dNTPs	NEB, Frankfurt, Germany
Dual-Luciferase Reporter Assay System	Promega, Madison, USA
EpiTect Bisulfite Kit	Qiagen, Hilden, Germany
Exo-Klenow-Fragment	Invitrogen, Karlsruhe
EZ DNA methylation kit	Zymo Research, Orange, USA
FastStart TaqDNA polymerase	Roche, Mannheim, germany
Gene expression hybridization Kit	Agilent, Waldbronn, Germany
HhaI Methylase	NEB, Frankfurt, Germany
Hpa II Methylase	NEB, Frankfurt, Germany
Human Cot-1 DNA	Invitrogen, Karlsruhe, Germany
Klenow Enzyme	NEB, Frankfurt, Germany
Klenow exo- (3'-5' exo minus)	NEB, Frankfurt, Germany
Lipofectamin transfection reagent	Invitrogen, Karlsruhe, Germany
Linear Amp. Kit plus, one colour	Agilent Technologies, Waldbronn, Germany
NucleoSpin Plasmid Quick Pure	Macherey-Nagel, Düren, Germany
NucleoSpin® Extract II	Macherey-Nagel, Düren, Germany

Material and equipment

Pepstatin	Roche, Mannheim, Germany
PicoGreen ds DNA Quantitation Reagent	MoBiTec, Göttingen
Plasmid Midi Kit	Qiagen, Hilden, Germany
PMSF (Phenylmethanesulfonylfluoride)	Sigma, Deisenhofen, Germany
Proteinase K	Roche, Mannheim
QIAquick PCR Purification Kit	Qiagen, Hilden, Germany
QuantiFast SYBR green	Qiagen, Hilden, Germany
Repli-G Midi Kit	Qiagen, Hilden, Germany
Restriction endonucleases	NEB, Frankfurt; Roche, Mannheim; Germany
Reverse Transkriptase SuperSkript II	Promega, Madison, USA
RNA 6000 Nano Kit	Agilent Technologies, Waldbronn, Germany
RNA Spike-in Kit	Agilent Technologies, Waldbronn, Germany
RNeasy Midi and Mini Kit	Qiagen, Hilden, Germany
S-adenosylmethionine (SAM)	NEB, Frankfurt, Germany
Shrimp Alkaline Phosphatase (SAP)	Sequenom, Hamburg, Germany
Sss I CpG methylases	NEB, Frankfurt, Germany
T-Cleavage MassCleave Reagent kit	Sequenom, Hamburg, Germany
TaqDNA Polymerase	Roche, Mannheim, Germany
T4 DNA Ligase	Promega, Madison, USA
T4 DNA Ligase buffer	NEB, Frankfurt, Germany
TOPO TA Cloning Kit	Invitrogen, Karlsruhe
Wizard DNA Clean-Up System	Promega, Madison, USA

3.5 Molecular weight standards

DNA ladder 1 kB Plus was purchased from Invitrogen (Karlsruhe, Germany). The Kaleidoscope Prestained standard protein marker was purchased from BioRad (Munich, Germany).

3.6 Oligonucleotides

3.6.1 Sequencing primers

Gene	Primer sequence (sense & antisense)
M13 reverse	5'-GGA AAC AGC TAT GAC CAT GAT-3'
T7	5'-TAA TAC GAC TCA CTA TA-3'

3.6.2 Real-time PCR primers for MClp

Gene	Primer sequence (sense & antisense)
CBX6	5'-AAGCTTCCGCCATTGCTCTG-3' 5'-TCCGTTCCTGGACAGCCC-3'
CDKN2B	5'-GGCTCAGCTTCATTACCCTCC-3' 5'-AAAGCCCGGAGCTAACGAC-3'
CHI3L1	5'-ATCACCCTAGTGGCTCTTCTGC-3' 5'-CTTTTATGGGAACTGAGCTATGTGTC-3'
COL14A1	5'-AGACGCAATGCAGTTCCATGG-3' 5'-ATCTCCCTACACCGTGAACCC-3'
CYP1B1	5'-TGTTGAATCCGTGCTTAGTAGAGACC-3' 5'-CAGAGTAGCATTCAGAAAGGCAGATGG-3'
CYP27B1	5'-CATCCGTTCTCTCTGGCTGTCC-3' 5'-CTGTCGAGGCTACACGAGCTGC-3'
Empty6.2	5'-GAAACCCTCACCCAGGAGATACAC-3' 5'-TGCAGTGGGACTTTATTCCATAGAAGAG-3'
DMRT2	5'-CACGTTTTTGCTAGAGGTGAGGG-3' 5'-TCCTCCATCCGTACTGACATAGGG-3'
ESR1	5'-GACTGCACTTGCTCCCGTC-3' 5'-AAGAGCACAGCCCGAGGTTAG-3'
FARP1	5'-GCTCCGTAGAGTTCCCGAAACC-3' 5'-AGCGAATCCCATGACAGTTCCC-3'
FNBP1	5'-ATCCAAAGGTCTGCACAAATGTTCCTG-3' 5'-CGAGGGAGAAAGATAAGCTGTGGG-3'
HOXD10	5'-TCTATAGTGACGCTACCTTTCCCG-3' 5'-CTTGAGAGGACAACGACATTTAGGG-3'
JUN	5'-AGGAGTTAGTGTGACAGGGTCGC-3' 5'-CCAAATCGCACTCTTATATCCTGGC-3'
JUN (p)	5'-ATTGGCTCGCGTCGCTCTC-3' 5'-GGAGCATTACCTCATCCCGTG-3'
KLF5	5'-AGACACTTCATTTAGTAGCTCTTTGGCG-3' 5'-GCCCTCTCACAGCAAGACCC-3'
KLF11	5'-GACAGCGGGCTAGATGTCTCC-3' 5'-GTCAGGGGAAGCCGAAACG-3'
KLF11 (p)	5'-GTTGAGGCCTCTAGGTGGGTCTC-3' 5'-CCACGCTTATAGGAACCTCCTGC-3'
LDLR	5'-GGGTACAAATAATCACTCCATCCCTG-3' 5'-TAAATCCCTCAGACTCCTCCCG-3'
MAFB	5'-TGTGCAGACTATGTATGGCTCCG-3' 5'-AAACACTCTGGGAGCCACAGG-3'
MAFB (p)	5'-TCGAGGTGTGTCTTCTGTTCGG-3' 5'-GACCTGCTCAAGTTCGACGTG-3'
MLF1	5'-AAATCTGATAGGCTTCATCCCATTTCC-3' 5'-GTCCTGTATCCGAAACATTCTCTGG-3'
PAX9	5'-CTCTGCTTGTCATAACTGCAACTCGG-3' 5'-TGATGACTGTGGATGGGAGGATAGG-3'
PDE4B	5'-CAGGAGGTCTGTGAGGTAGGTG-3' 5'-TGTGTAGTAGGTTGTAACTGCTGAGG-3'

Material and equipment

Gene	Primer sequence (sense & antisense)
PFC	5'-CGTTACGGGTTTCCTGATTGGC-3' 5'-GGAATCTAGGGAGGTCCAGGAG-3'
PLA2G7	5'-GTGCTGGTGTCATTTCTCCCTG-3' 5'-TCTAGCTCCATTTCTCCTCAGACC-3'
RAB3C	5'-TGAGGGATCGGGCTATTCGC-3' 5'-GCCAAGAGAGGAGATCAATGCC-3'
RAX	5'-CATGGACACCCGTGAATTCCGAG-3' 5'-AGGTAAAGCGCCCAGGTTGAG-3'
RGMA	5'-AAAGACCGTATCGCACTCCCTC-3' 5'-CGCAGAGACTGGAAAGAACCG-3'
RPIB9	5'-AAAGACTCTACACTGGCACCACG-3' 5'-TAGTGCCGACATTTCTTGCCC-3'
RPP30	5'-AGCTTCTAAGTTACTATCAGCCCTTCC-3' 5'-GTATTGTTCCAACACTCCCACGTCC-3'
SETBP1	5'-TGTGCGTTTCTAGAGGAGCCG-3' 5'-AAATCGATACCGAAGGGTTCCC-3'
SLITRK3	5'-TACCTCTTACAACACCAGCGAGC-3' 5'-GGATCAGTTAGGTGTAAGGACGTTGG-3'
SNRPN	5'-TACATCAGGGTGATTGCAGTTCC-3' 5'-TACCGATCACTTCACGTACCTTCG-3'
SSIAH2	5'-CTGAGACACTCCGCTCCAGC-3' 5'-TGTTATTGGCTGTCTCTGCACCTC-3'
TGIF	5'-GTCCGGGAAGGAACTGTGCTC-3' 5'-CTGCTCGGGACAGAAGAGAACAC-3'
TLR2	5'-TGTGTTTCAGGTGATGTGAGGTC-3' 5'-CGAATCGAGACGCTAGAGGC-3'
ZFP36L1	5'-AAACATTGTCCCGAGACTCACTTCC-3' 5'-GTCTGTCCAGCGGCATTACC-3'
ZNF516	5'-CAGGTGATGATGGAACCCACTC-3' 5'-TGCTGCCCTTCACTTTTCTACG-3'
ZNF516 (p)	5'-CCCTCAGTGTGGCAGAACTTTG-3' 5'-CCCAGCCTGGAAATGGTC-3'

3.6.3 Real-time PCR primers for ChIP-on-chip

Gene	Primer sequence (sense & antisense)
HDAC3	5'-TCAGCTCTCCCGGTATCTGG-3' 5'-GACAAATGGCCCTCGCATCC-3'
LDHB	5'-GTCGTGCGGAGAAGACAAAGTCAG-3' 5'-CTAAGAGGCTGCGGTGGTTGTG-3'
CTCF	5'-GTCCCTTCCCTTATCAGCACCC-3' 5'-GCACGGTTTAATCGCTCCACAG-3'
RAN	5'-CGTCTCCGGCGTTTGAATTGC-3' 5'-GCGATACCTTCCAGAAGCGTC-3'
CAPNS1_1	5'-AGACCTGGATCCAGCTAGCC-3' 5'-AACTCTCGGGTCGGACACTG-3'

Gene	Primer sequence (sense & antisense)
CAPNS1_2	5'-CAATGTCCGCTTCGGCTCTAGG-3' 5'-TGACTCAGGCCGCAACTCTC-3'
CXCR4	5'-AGATGCGGTGGCTACTGGAG-3' 5'-CAGAAACTTCAGTTTGTTGGCTGCG-3'
MED8	5'-CTAGTACGCCCAACGCAACTC-3' 5'-CTGATTGGTGGACGAAGCCTTCTC-3'
LOC116349	5'-TAGCCGTAGAGGGTGAGTCG-3' 5'-ACAGGGAGAGCAAGGATGAAAGAC-3'
AIP	5'-CGCAGAGAACCAATCACCATCC-3' 5'-CTTCGGCAACTCCTAGCACC-3'
KLF15	5'-ACCTCCTTGCTTCCCACCTC-3´ 5´-CAGGCCAGTCTCACGTTCTCAC-3´

3.6.4 Real-time RT-PCR primer

Gene	Primer sequence (sense & antisense)
MAFB	5'-GAGCCGGAGAGAGAGACG-3' 5'-AGGAGTCTCCAGATGGCCTTG-3'
JUN	5'-CGGCGGTAAAGACCAGAAGG-3' 5'-CGCCCAAGTTCAACAACCG-3'
KLF11	5'-ACCTACTTCAAAAGTTCCCACC-3' 5'-CATGAAACGTCGGTCACACAC-3'
SSIAH2	5'-GTTTCAGCACTACAAGGCTAAACGG-3' 5'-AAGCTGCCTTGCTCTGGAGC-3'
ZNF516	5'-GTTCTGAAGTTCATACCACCTCCG-3' 5'-TCAGAGGCACTGTCTGGACGG-3'

3.6.5 LM-PCR oligonucleotides

Gene	Primer sequence (sense & antisense)
LM_JW102_sticky	5'-GCGGTGACCCGGGAGATCTGAATTCT-3'
LM_JW103_sticky	5'-GAATTCAGATC-3'

3.6.6 Bisulfite amplicon generation (Nested PCR)

Gene		Primer sequence (sense & antisense)
RAB3C	outer	5'-ATTGGGAGAGGTAATTTAGGAG-3' 5'-ATTTTAAACAAACACTCTTATCCTC-3'
	inner	5'-TTTGGAAAGGAGTAGGGAGG-3' 5'-ATCCCTCATCAAAACAACCC-3'
MAFB	outer	5'-GGGTAYGGYGTGGTATTGGG-3' 5'-TAAAATAAATCACAACTTAACCTATCCATC-3'

Material and equipment

Gene		Primer sequence (sense & antisense)
JUN	inner	5'-TTTAATTTAATTTTGTGGGGTGGT-3' 5'-CTCTAAAAACCACAAATCTCTTAAAACC-3'
	outer	5'-TTTAGATGGGAATAAGYGTGTAGG-3' 5'-TACTACAAATCCAACTTCAAACC-3'
SSIAH2	inner	5'-TYGGGAAAATAAGTTTAGAAGG-3' 5'-ACTCTTATATCCTAACATCCTATCC-3'
	outer	5'-TTTAATATATGGGATAGAGAGAATTTGG-3' 5'-TTCTATCCTTTTAATTAACCRCCTCAC-3'
KLF11	inner	5'-AAATAGTAGGGGGAGTGATGGG-3' 5'-AAACCCAAAAACTCACAACTTCC-3'
	outer	5'-TGTTTATGTGAGTGGTGGGG-3' 5'-ACCCACCTAAAAACCTCAACC-3'
ZNF516	inner	5'-TTGTTTTYGTTTTTTGGATGGAG-3' 5'-TATTTTTAACTTCTATCATTCTCCC-3'
	outer	5'-CACCCAATTCTACCCCTCC-3' 5'-ATTTTTTTATTGGGAGTTGATG-3'
CPM	inner	5'-ACCTCTCCATTACATCATCCC-3' 5'-GTTTTTGGTAAATTTTAGAAGGTG-3'
	outer	5'-TTGGTTAGTTAGTTGGGTTTTGG-3' 5'-AAACAATTATACTAACCTTCTTCTCTTTCC-3'
	inner	5'-TTGGTATTTAGATTTGGAGTGGG-3' 5'-TAATATACAATAACTTCCACCATAACCA-3'

3.6.7 MassARRAY QGE

3.6.7.1 Oligonucleotides

Gene	Primer sequence (sense & antisense)
MGMT_MBD-Fc_1_TAG	5'- ACG TTG GAT GCG CCC CTA GAA CGC TTT G -3' 5'- ACG TTG GAT GAG ACA CTC ACC AAG TCG CAA AC -3'

3.6.7.2 Competitors

Gene	Primer sequence (sense & antisense)
Comp_MGMT-1	5'- CGC CCC TAG AAC GCT TTG CGT CCC GAC GCC CGC AGG TCC CCG CGG TGC GCA CCG TTT GCG ACT TGG TGA GTG TCT 5'- AGA CAC TCA CCA AGT CGC AAA CGG TGC GCA CCG CGG GGA CCT GCG GGC GTC GGG ACG CAA AGC GTT CTA GGG GCC

3.6.8 Bisulfite amplicon generation (MassARRAY)

All primers designed for methylation analysis using the MassCLEAVE assay are or will be available within the supplementary information of the corresponding publications.

3.7 Antibodies

Chromatin immunoprecipitation (ChIP)

Anti-YY1	Santa Cruz, sc-1703
Anti-NRF1	Abcam, ab34682
Anti-Sp1	Upstate, 07645
Anti-rabbit IgG	Upstate

Western Blot

Goat anti-IgG F(c), HRP conjugated	Rockland, Gilbertsville, USA

3.8 Antibiotics

Ampicillin	Ratiopharm, Ulm, Germany
Hygromycin	Clontech, Mountain View, USA
Zeozin	Invitrogen, Karlsruhe, Germany

3.9 Plasmids

pCpG-mcs	Invivogen, San Diego, USA
pGL3-Basic	Promega, Mannheim, Germany
pCR®2.1-TOPO	Invitrogen, Karlsruhe, Germany

3.10 *E.coli* strains

PIR1	F-Δlac169 rpoS(Am) robA1 creC510 hsdR514 endA recA1 uidA(ΔMluI)::pir-116
TOP10	F-mcrA Δ(mrr-hsdRMS-mcrBC) Φ80lacZΔM15 ΔlacX74 recA1 deoR araD139 Δ(ara-leu)7697 galU galK rpsL (StrR) endA1 nupG

Material and equipment

DH10ß   F-mcrA Δ(mrr-hsdRMS-mcrBC) Φ80lacZΔM15 ΔlacX74 recA1 deoR araD139 Δ(ara-leu)7697 galU galK rpsL endA1 nupG

3.11 Cell lines

Human cell lines

THP-1	Human acute monocytic leukemia (DSMZ no ACC 16)
U937	Human histiocytic lymphoma (DSZM no ACC 5)
KG-1	Human acute myeloid leukemia (DSZM no ACC 14)

Insect cell lines

Drosophila Schneider 2 (S2) cells

3.12 Databases and software

Agilent feature extraction 9.5.1	Agilent Technologies, Waldbronn, Germany
BLAT	http://genome.brc.mcw.edu
EpiTYPER 1.0.5	Sequenom, Hamburg, Germany
Generunner version 3.05	Agilent Technologies, Waldbronn, Germany
Genespring 10.0.2	
Perlprimer version 1.1.14	
PubMed	www.ncbi.nlm.nih.gov/entrez
Spotfire decision site 7.0	
UCSC Genome. Browser	www.genome.ucsc.edu

Reference sequence: Genomic locations are based on the March 2006 human reference sequence (NCBI Build 36.1) that was produced by the International Human Genome Sequencing Consortium.

The following databases were used to annotate genes associated with CpG islands that were either bound by general transcription factors (Sp1, NRF1 or YY1) or associated with a particular methylation status:

Biological Process:	Functional groupings of proteins (Gene Ontology, http://www.geneontology.org/)
Molecular Function:	Mechanistic actions of proteins (Gene Ontology)
Cellular Component:	Protein localization (Gene Ontology)
KEGG Pathways:	Groups of proteins in the same pathways (From KEGG, http://www.genome.jp/kegg/pathway.html)
Interactions:	Groups of proteins interacting with the same protein (From NCBI Gene, http://www.ncbi.nlm.nih.gov/sites/entrez?db=gene)
Interpro:	Proteins with similar domains and features (Interpro, http://www.ebi.ac.uk/interpro/)
Pfam:	Proteins with similar domains and features (Pfam, http://pfam.sanger.ac.uk/)
SMART:	Proteins with similar domains and features (SMART, http://smart.embl-heidelberg.de/)
Gene3D:	Proteins with similar domains and features (Gene3D Database, http://gene3d.biochem.ucl.ac.uk/Gene3D/)
Prosite:	Proteins with similar domains and features (Prosite Database, http://ca.expasy.org/prosite/)
PRINTS:	Proteins with similar domains and features (PRINTS Database, http://www.bioinf.manchester.ac.uk/dbbrowser/PRINTS/index.php)
Chromosome Location:	Genes with similar chromosome localization
miRNA Targets:	Genes targeted by similar miRNAs (miRBase target database http://microrna.sanger.ac.uk/)

3.13 Statistical testing

All statistical analysis of enrichment data (motifs or attributes) was performed using a cumulative hypergeometric distribution (or Fisher Exact test, referred to as the hypergeometric test). Statistical testing of differences in mRNA level distributions was done using the two-sided Mann–Whitney U test.

4 Methods

4.1 General cell culture methods

For washing and harvesting, mammalian cells were centrifuged using the general cell program: 8 min, 300×g, 4 °C.

4.1.1 Cell line culture conditions and passaging

Cells were cultured in RPMI 1640 (HyClone) or DMEM (Gibco) routinely supplemented with 10% inactivated FCS, L-glutamine (2 mM), sodium pyruvate (1 mM), antibiotics (50 U/ml penicillin and 50 µg/ml streptomycin), 2 ml vitamins, non essential amino acids and 50 µM ß-mercaptoethanol. Media supplements were purchased from Gibco and Biochrome (L-glutamine), respectively.

FCS was heat inactivated for 30 min at 56 °C before use. Exceeding incubation times and higher temperatures should be avoided because heat sensitive growth factors could be damaged. Each batch of FCS as well as each RPMI batch was tested before use.

Culturing of cells was performed at 37 °C, with 5% CO_2 and 95% relative humidity in an incubator. U937, THP-1 and KG-1 cells grow in suspension and were split 1:4 to 1:8 in fresh medium every 2-4 days.

4.1.2 Culturing of stably transfected *Drosophila* S2 cells and expression of the methyl binding polypeptide MBD-Fc

MBD-Fc stands for a fusion protein consisting of the methyl-CpG binding domain (MBD) of human MBD2 (methyl-CpG binding domain protein 2) and the Fc-tail of human IgG1. The MBD-Fc vector was stably transfected into *Drosophila* S2 cells using the Effectene transfection reagent (Qiagen) and hygromycin selection. A detailed description of design and generation of the fusion protein can be found in Gebhard, 2005 and Gebhard et al., 2006b.

Expansion in cell culture bottles

MBD-Fc S2 cells were seeded at a density of $1-2\times10^6$ cells/ml in Insect-Xpress medium (Lonza) including 50 U/ml penicillin and 50 µg/ml streptomycin but without FCS at 21-23 °C. 400 µg hygromycin was added for selection of plasmid containing cells. Cells were splitted once a week, without exceeding 10×10^6 cells/ml.

Protein production

Cells were transferred into 2,000 ml roller bottles and cultured at a density of 4×10^6 cells/ml in up to 400 ml Insect-Xpress medium supplemented with penicillin, streptomycin and hygromycin as described above. Cells should not exceed a density of 10×10^6 cells/ml. For large-scale protein production, after 3-5 days the culture media was exchanged and 5×10^6 cells/ml were seeded in 400 ml Insect-Xpress medium. Instead of hygromycin, 0.5 mM $CuSO_4$ was added to stimulate the metal inducible promoter of the used vector. The MBD-Fc containing culture medium was harvested after 4 days like described in section 4.2.1. For recovery, cells were cultured again in Insect-Xpress medium containing standard antibiotics and selection antibiotic for 3-5 days. The cycle of production was repeated until protein quality and amount clearly decreased.

4.1.3 Assessing cell number and vitality

The number of viable and dead cells was determined by Trypan blue exclusion. Cell suspensions were diluted with Trypan blue solution and then counted in a Neubauer haemocytometer. Dead cells appear blue since the blue stain is able to enter the cytoplasm. The concentration of viable cells was then calculated using the following equation:

Number of viable cells/ml $C = N \times D \times 10^4$

    N: average of unstained cells per corner square (1 mm containing 16 sub-squares)
    D: dilution factor

Required solutions and materials:

    Trypan blue solution: 0.2% (w/v)   Trypan blue in 0.9% NaCl solution
    Neubauer haemocytometer slide with coverslip

4.1.4 Freezing and thawing cells

Cells were harvested and resuspended at $5\text{-}10 \times 10^6$ cells/ml in 800 µl ice cold medium, including 10% FCS. After inverting the mix and transferring it into cryo-vials, 160 µl DMSO (10% final) and 640 µl FCS (40% final) were added and the tubes were rapidly inverted to mix cells properly. To allow gradual freezing at a rate of 1°C/min, the cryo-vials were placed in isopropanol-filled cryo-containers (Nalgene) for two hours, then transferred to -80°C for 48 h. For long-term storage, samples were transferred in liquid nitrogen (-196°C).

4.1.5 Mycoplasma assay

Cell lines were routinely checked for mycoplasma contamination by the MycoAlert® Mycoplasma detection assay (Cambre, Rockland, USA) according to the manufacturer's instructions.

4.1.6 Isolation of human monocytes

Peripheral blood mononuclear cells (PB-MNCs) were separated by leukapheresis of healthy donors (Graw et al., 1971), followed by density gradient centrifugation over Ficoll/Hypaque (Johnson et al., 1977). Monocytes were then isolated from MNCs by counter current centrifugal elutriation (Sanderson et al., 1977).

Elutriation was performed in a J6M-E centrifuge equipped with a JE 5.0 elutriation rotor and a 50 ml flow chamber (Beckman, Munich, Germany). After sterilizing the system with 6% H_2O_2 for 20 min, the system was washed with PBS. Following calibration at 2,500 rpm and 4°C with Hanks BSS, MNCs were loaded at a flow rate of 52 ml/min. Fractions were collected and the flow-through rate was sequentially increased according to Table 4-1.

Table 4-1 Elutriation parameter and cell types

Fraction	Volume (ml)	Flow rate (ml/min)	Main cell type contained
Ia	1000	52	platelets
Ib	1000	57	
IIa	1000	64	
IIb	500	74	B- and T- lymphocytes, NK cells
IIc	400	82	
IId	400	92	
III	800	130	monocytes

Monocytes represent the largest cells within the MNCs and are therefore mainly obtained in the last fraction. Monocytes were >85% pure as determined by morphology and CD14 antigen expression. Low amounts of monocytes may be also detected in the IId fraction. Monocytes (fraction III) were centrifuged (8 min, 300×g, 4°C), resuspended in RPMI culture medium and counted. Monocyte yields were donor-dependent, typically between 10-20% of total MNCs. Supernatants of monocyte cultures were routinely collected and analyzed for the presence of interleukin-6 (IL-6), which was usually low, indicating that monocytes were not activated before or during elutriation.

4.2 General protein biochemical methods

4.2.1 Purification of the recombinant protein MBD-Fc

4.2.1.1 Dialysis

The MBD-Fc containing culture supernatant (see section 4.1.2) was harvested by centrifugation of the cells at 320×g for 10 min at 4°C. To remove (dead) cells and debris the supernatant was centrifuged at 2,000×g for 20 min at 4°C before the final centrifugation step of 15,000×g for 1 hour at 4°C to get rid of smaller debris. Afterwards the supernatant was dialyzed against 1×TBS (pH 7.4) for 3-4 days whereas the buffer was exchanged twice a day.

Required buffers:

10×TBS pH 7.4	151.4 g	(500 mM)	Tris
	219.2 g	(1.5 M)	NaCl
	9.3 g	(10 mM)	EDTA
	125 mg	(0.05%)	NaN_3
	Add ddH_2O to 2500 ml		

4.2.1.2 Affinity chromatography

After dialysis the protein containing supernatant was purified and enriched using a ProteinA sepharose column (Amersham):

The column (Pharmacia) was filled with 3 ml rProteinA sepharose beads (GE Healthcare, Uppsala, Sweden) in 1×TBS. After washing the column with 1×TBS, the dialyzed protein supernatant was loaded, followed by another washing step with 1×TBS. Constant flow rate of dialyzed cell culture supernatant was achieved using a peristaltic pump (Heidolph, Schwabach, Germany). Elution was performed in 1 ml fractions with elution buffer into 1.5 ml tubes each containing 50 µl neutralization buffer. The different fractions were measured using a Biophotometer (Eppendorf, Hamburg, Germany). The protein containing fractions (determined by a photometer) were combined and dialyzed against 1×TBS (as described above) using Slide-A-Lyzer Dialysis Cassettes (Pierce, Rockford, USA).

Regeneration was performed by washing the column with 3 M KCl and finally with 1×TBS, now prepared for another purification cycle or for storage at 4°C.

Required buffers and solutions:

Elution buffer pH 3.0	2.9 g	(0.1 M)	Citric acid
	Add ddH$_2$O to 100 ml		
Neutralization buffer pH 8.8	18 g	(1.5 M)	Tris
	Add ddH$_2$O to 100 ml		
Recovering solution	22.4 g	(3.0 M)	KCl
	Add ddH$_2$O to 100 ml		

4.2.1.3 Conservation of the purified MBD-Fc

To stabilize and preserve the protein, 0.2% gelatine and 0.05% NaN$_3$ were added. The MBD-Fc fusion protein was now ready for further experiments or for long-term storage at 4°C.

4.2.1.4 Quantification and quality control of MBD-Fc

Quality of each protein batch was assessed by SDS-PAGE (see section 4.2.2) followed by Coomassie staining (or Western Blot analysis as described in section 4.2.3) as well as by control-MCIp (see section 4.4.4).

Protein concentration was determined relative to a BSA standard curve using a densitometer after SDS-PAGE.

4.2.2 Discontinuous SDS-PAGE

Protein samples were separated by using a discontinuous gel system. This technique separates proteins according to their electrophoretic mobility, which is besides other characteristics a function of the polypeptide chain length. A polyacrylamide gel is composed of stacking and separating gel layers that differ in salt and acrylamide (AA) concentration. For a sodium dodecyl sulfate (SDS) polyacrylamide (AA) gel electrophoresis (SDS-PAGE) the protein preparation was diluted 1:5 with H_2O in a volume of 10 µl and supplemented with 10 µl SDS sample buffer. Accordingly, a bovine serum albumin (BSA) standard curve was prepared containing four different dilutions comprising 1, 0.5, 0.25 and 0.125 mg/ml. All samples were incubated to 95°C for 5 minutes and subsequently loaded into a SDS-PAGE assembly together with a pre-stained protein size standard (Bio-Rad Laboratories, Munich, Germany).

Table 4-2 SDS-PAGE stock solutions

Stock solution	Separating gel	Stacking gel
Final AA concentration	13.5%	5%
Stacking gel buffer	-	25 ml
Separating gel buffer	25 ml	
SDS	1 ml	1 ml
Rotiphorese Gel 30 (30%)	45 ml	16.65 ml
H_2O	adjust to 100 ml	

Table 4-3 SDS-PAGE gel mixture

Stock solution	Separating gel	Stacking gel
Separating gel stock solution	10 ml	-
Stacking gel stock solution	-	5 ml
TEMED	10 µl	5 µl
Ammoniumpersulfate 10% freshly prepared	50 µl	40 µl

The separating gel was prepared the day before electrophoresis and overlaid with water-saturated isobutanol until it was polymerized. Isobutanol was exchanged for separating gel buffer diluted 1:3 with water and the gel was stored overnight at 4°C. The following day, the stacking gel was poured on top of the separating gel, and the comb was inserted immediately. After polymerization, the gel was mounted in the electrophoresis tank, which was filled with 1×Laemmli buffer. Protein samples were loaded and the gel was run with 25 mA/110 volts until the sample buffer bands reached the surface of the stacking gel. Then the voltage was increased to 200 V and the gel was run for 2-4 h. Proteins were then resolved through the separating gel according to their size.

Methods

Required buffers and solutions:

Separating gel buffer	90.83 g	(1.5 M)	Tris/HCl, pH 8.8
	Add ddH$_2$O to 500 ml		
Stacking gel buffer	30 g	(0.5 M)	Tris/HCl, pH 8.8
	Add ddH$_2$O to 500 ml		
SDS (10%)	10 g	(10%)	SDS
	Add ddH$_2$O to 500 ml		
Ammonium persulfate (APS)	100 mg	(10%)	Ammonium persulfate
	Add ddH$_2$O to 1 ml		
Laemmli buffer (5×)	15 g	(40 mM)	Tris
	21 g	(0.95 M)	Glycine
	15 g	(0.5%)	SDS
	Add ddH$_2$O to 3000 ml		

4.2.3 Western Blot analysis and immunostaining

After separation by SDS-PAGE, proteins were blotted electrophoretically onto a PVDF membrane (Immobilon-P, Millipore) using a three-buffer semi-dry system and visualized by immunostaining using specific antibodies and the ECL detection kit.

The membrane was cut to gel size, moistened first with methanol followed with buffer B and placed on top of three Whatman 3MM filter paper soaked with buffer A (bottom, on the anode) followed by three Whatman 3MM filter paper soaked with buffer B. The SDS-PAGE gel was then removed from the glass plates, immersed in buffer B and placed on top of the membrane. Another three Whatman 3MM filter papers soaked with buffer C were placed on top of the gel followed by the cathode. Air bubbles in between the layers had to be avoided. Protein transfer was conducted for 30 – 45 min at 0.8 mA/cm^2 gel surface area.

Required buffers:

Buffer A	36.3 g	(0.3 M)	Tris, pH 10.4
	200 ml	(20%)	Methanol
	Add ddH$_2$O to 1000 ml		
Buffer B	3.03 g	(25 mM)	Tris, pH 10.4
	200 ml	(20%)	Methanol
	Add ddH$_2$O to 1000 ml		
Buffer C	5.2 g	(4 mM)	ε-amino-n-caproic acid, pH 7.6
	200 ml	(20%)	Methanol
	Add ddH$_2$O to 1000 ml		

Blotted membranes were then blocked either with 5% milk in PBST for 1 h at RT then washed once for 5 min with PBST or TBST before incubation at RT for 1 h with the primary antibody. After washing three times 10 min with the appropriate washing buffer, the membrane was incubated for 1 h at RT with a horseradish-peroxidase (HRP)-coupled secondary antibody, detecting the isotype of the first antibody. Three washing steps of 3×10 min preceded the visualization of bound antibody using the ECL kit. Blots were exposed to an autoradiography film (Hyperfilm^{TM} ECL, Amersham) for 5 seconds to 30 min depending on the signal intensity.

Required buffers and materials:

TBS (2×)	9.16 g	(20 mM)	Tris /HCl, pH 7.4
	35.1 g	(150 mM)	NaCl
	Add ddH$_2$O to 2000 ml		
TBST (1×)	500 ml	(1×)	TBS (2×)
	1 ml	(0.1%)	Tween 20
	Add ddH$_2$O to 1000 ml		

4.2.4 Coomassie staining of SDS gels

SDS-gels were tossed in ddH$_2$O (three times, 5 min each) and subsequently incubated in the Coomassie solution for about 20 – 60 min. After washing overnight in ddH$_2$O, proteins appear as blue bands on a transparent background. For documentation purposes the ready stained gel was scanned using a personal Densitometer SI (Molecular Dynamics). The gel image was loaded into the ImageQuant 5.0 software and protein bands were quantified using the BSA-standard curve as a reference.

Required solution:
    Coomassie Bio Safe   Bio Rad, Munich, Germany

4.3 General molecular biological methods

4.3.1 Bacterial culture

4.3.1.1 Bacterial growth medium

E.coli strains were streaked out on solid LB-agar with specific antibiotics and grown overnight (O/N) at 37 °C. Single colonies were then picked into liquid LB-medium and then incubated O/N with shaking at 200 rpm.

LB-medium:	10 g	NaCl
	10 g	Bacto Tryptone (Difco)
	5 g	Yeast extract
	Add ddH$_2$O to 1000 ml, autoclave	
LB-agar plates:	15 g	Agar
	10 g	NaCl
	10 g	Bacto Tryptone (Difco)
	5 g	Yeast extract
	Add ddH$_2$O to 1 l, autoclave, cool to 50 °C and add the appropiate antibiotic.	
	Pour the agar solution into 10 cm Petri dishes and store inverted at 4 °C.	

4.3.1.2 Transformation of chemically competent *E.coli*

Chemically competent *E.coli* (50 µl) were thawed on ice, 1-25 ng plasmid DNA in 2-5 µl volume (e.g. 2 µl from TOPO-cloning) was added and the suspension was mixed gently and incubated on ice for 30 min. Cells were then heat-shocked in a water bath at 42 °C for 30 s, immediately cooled on ice for 2 min and 250 µl SOC medium was added. To express the resistance, bacteria were incubated for 1 h at 37 °C with shaking. For blue/white screening of insert-containing clones after transformation (in case of TOP10 cells), 40 µl of X-gal was dispersed on a prewarmed LB-plate prior to use and incubated at 37 °C for additional 30 min. Afterwards 50-150 µl of the transformation reaction were plated and incubated at 37 °C on LB-agar containing the antibiotic necessary for selection of transformed cells overnight.

SOC medium	20 g	(2%)	BactoTrypton (Difco)
	5 g	(0.5%)	BactoYeastExtract (Difco)
	0.6 g	(10 mM)	NaCl
	0.2 g	(3 mM)	KCl
	Add ddH$_2$O to 1000 ml, autoclave and add to the cooled solution:		

		10 ml (10 mM) 10 ml (10 mM) 10 ml (20 mM)	MgCl$_2$ (1 M), sterile filtered MgSO$_4$ (1 M), sterile filtered Glucose (2 M), sterile filtered
X-gal		40 mg	X-gal (5-bromo-4-chloro-3-indolyl-ß-D-galactoside) In 1 ml DMF, store at -20 °C protected from light

4.3.1.3 Glycerol stock

For long-term storage, bacteria were stored in 20% glycerol by adding 500 µl liquid culture to 200 µl of 80% glycerol, mixing and freezing at -80 °C.

4.3.1.4 Plasmid isolation from *E.coli*

To check if the isolated single *E.coli* colonies contained the correct plasmid, a DNA mini-prep was carried out using NucleoSpin® Plasmid Quick Pure Kit from Macherey-Nagel following the supplied instructions. Afterwards the plasmid constructs were sequenced. To isolate larger amounts of ultra pure DNA (100 µg) for transfection experiments, plasmids were isolated using the QIAGEN Plasmid Midi Kit for endotoxin-free midi-preps according to the manufacturer's instructions.

4.3.2 Molecular cloning

Direct cloning of PCR products was done using the TOPO-TA Cloning kit (Invitrogen) according to the manufacturer's instructions. Alternatively, DNA fragments were PEG-precipitated and the precipitates as well as the cloning vector were digested with the appropriate restriction endonucleases (New England Biolabs or Roche). For directional cloning, restriction sites were introduced by adding the appropriate recognition sequences to the primer sequences. The cut fragments and the vector were gel-purified and combined in a 10 µl ligation reaction with a 3- to 5-fold molar excess of insert to vector, using 25-50 ng of vector. Ligation was carried out overnight at 16 °C with 1 U T4 DNA Ligase or alternatively 5 min using a rapid ligation system. 2 µl of the reaction was then used to transform chemically competent *E.coli* (see section 4.3.1.2).

Successful insertion of the fragment into the vector was controlled by preparing plasmid DNA from liquid cultures (see section 4.3.1.4). To control correct insertion and sequence integrity, plasmid constructs were sequenced using vector-specific primers.

4.3.2.1 PEG precipitation

To precipitate DNA from small volumes, e.g. PCR reactions or endonuclease digestion, one volume of PEG-mix was added to the DNA-containing solution, vortexed and incubated for 10 min at RT. After centrifugation (10 min, 13,000 rpm, RT), the supernatant was discarded and the precipitated DNA was washed by carefully adding 200 µl 100% EtOH to the tube wall opposite of the pellet, followed by a centrifugation step (10 min, 13000 rpm, RT) and careful removal of the supernatant. The pellet was dried and resuspended in H_2O at half to three-quarters of the initial volume.

PEG-mix	26.2 g	(26.2%)	PEG 8000
	20 ml	(0.67 M)	NaOAc (3 M) pH 5.2
	660 µl	(0.67 mM)	$MgCl_2$ (1 M)
	Add ddH_2O to 250 ml		

4.3.2.2 Restriction endonuclease digestion

To verify the presence and orientation of plasmid insert, or to clone insert DNA into a plasmid, DNA was digested with appropriate restriction enzymes. Enzymes and their buffers were purchased from Roche or New England Biolabs (Germany). The digestion of plasmid DNA or PCR products was carried out using 10 U enzyme/1 µg DNA in 20 µl at 37°C for 2 hours. Digestion of genomic DNA was performed overnight with 1.5 U/µg DNA in 30 µl reaction volume (see section 4.3.3.4).

4.3.2.3 CIAP treatment

To prevent re-ligation of digested vectors, vector-ends were treated with CIAP (calf intestinal alkaline phosphatase, Roche) at 37°C for 30 min before gel extraction.

4.3.2.4 Gel purification

To purify DNA from analytical agarose gels, desired bands were excised under UV illumination and purified with the QIAEX II gel extraction kit (Qiagen) or NucleoSpin Extract II following the manufacturer's instructions.

4.3.2.5 Ligation reaction

Restriction enzyme treated vectors and PCR products were ligated in a 10 µl reaction at a 3- to 5-fold molar excess of insert to vector, using 25-50 ng of vector. Ligation was carried out overnight at 16°C with 1U T4 DNA ligase and 1 µl T4 DNA ligase buffer.

4.3.2.6 Sequencing

DNA sequencing was done by Geneart (Regensburg, Germany) with ABI sequencing technology based on the Sanger didesoxy method. Sequence files were analyzed and aligned with Generunner, Bioedit or with the Blat function of the UCSC genome browser.

4.3.3 Preparation and analysis of DNA

4.3.3.1 DNA preparation from normal cells

Genomic DNA (gDNA) was isolated using the Qiagen Blood & Cell Culture DNA Midi Kit or, for smaller cell numbers, the Blood and Tissue Culture Kit (Qiagen). gDNA concentration was then determined with the NanoDrop spectrophotometer and quality was assessed by agarose gel electrophoresis.

4.3.3.2 DNA preparation from clinical samples

Colorectal cancer samples were collected from 20 patients who underwent colon resection for biopsy-proven invasive colorectal adenocarcinoma at the University of Regensburg. The study was performed in agreement with the Institutional ethical review board of the University of Regensburg (05/003). The tissue was snap frozen and stored at −80°C (in cooperation with PD. Dr. W. Dietmeier, Department of Pathology). Each resection specimen was staged and graded by routine pathology analysis according to the TNM classification by the American Joint Committee on Cancer. DNA from frozen colon tissues was isolated using the PUREGENE™ DNA Purification Kit (Gentra, Minneapolis, USA) according to the supplier's recommendation.

Leukemic blasts and bone marrow cells from AML patients were collected during routine diagnostic bone marrow aspirations (in cooperation with Prof. Dr. G. Ehninger, TU Dresden). Patients had given informed consent to additional sample collection and analyses according to a protocol approved by the local ethical committee.

4.3.3.3 Agarose gel electrophoresis

The required amount of agarose as determined according to Table 4-4 was added to the corresponding amount of 1×TAE. The slurry was heated in a microwave oven until the agarose was completely dissolved. The ethidium bromide was added after cooling the solution to 50-60 °C. The gel was cast and mounted in the electrophoresis tank and covered with 1×TAE. DNA-containing samples were diluted 4:1 with DNA loading dye (5×), mixed and loaded into the slots of the submerged gel. Depending on the size and the desired resolution, gels were run at 40-100 volts for 30 min to 3 h.

Table 4-4 Agarose concentration for different separation ranges

Efficient range of separation (kb)	% agarose in gel
0.1 – 2	2.0
0.2 – 3	1.5
0.4 – 6	1.2
0.5 – 7	0.9
0.8 - 10	0.7
genomic DNA	0.5

Required buffers:

TAE (50×)	252.3 g	(2 M)	Tris	
	20.5 g	(250 mM)	NaOAc/HOAc, pH 7.8	
	18.5 g	(50 mM)	EDTA	
	Add ddH$_2$O to 1 l			
EDTA (0.5 M)	18.6 g	(0.5 M)	EDTA/NaOH, pH 8.0	
	Add H$_2$O to 100 ml			
DNA loading dye	500 µl	(50 mM)	Tris/HCl, pH 7.8	
DNA-LD (5×)	500 µl	(1%)	SDS (20%)	
	1 ml	(50 mM)	EDTA (0.5 M), pH 8.0	
	4 ml	(40%)	Glycerol	
	10 mg	(1%)	Bromophenol blue	
	Add H$_2$O to 10 ml, store at 4 °C			
1.0% Agarose	1 g	(1%)	Agarose (Biozym)	

Add 1×TAE to 100 ml and heat in a microwave until agarose is completely dissolved.
Cool to 50 °C and add 2.5 µl ethidium bromide (10 mg/ml) (Sigma).

4.3.3.4 Restriction endonuclease digestion

Digestion of genomic DNA was performed overnight with 1.5 U/µg DNA in 30 µl reaction volume.

4.3.3.5 Quantification of DNA

The exact DNA concentration was determined either by using the PicoGreen dsDNA Quantitation Reagent (Molecular Probes) or by using the NanoDrop spectrophotometer.

4.3.4 Polymerase chain reaction (PCR)

The polymerase chain reaction (PCR) allows *in vitro* synthesis of large amounts of DNA by sequence-specific polymerization of nucleotide triphosphates catalyzed by DNA polymerase (Mullis et al., 1986). The polymerization reaction is "primed" with small oligonucleotides that anneal to the template DNA strand through base pairing, giving the reaction its specificity by defining the borders of the segment to be amplified. Standard applications of PCR reactions are explained in the following and are used unless otherwise mentioned. More specialized applications are explained in more detail within the specific method.

4.3.4.1 Primer design

Unless otherwise mentioned sequences for generating primers were extracted using the UCSC Genome Browser. In general primers were designed using PerlPrimer Software and controlled using PCR and BLAT functions of the UCSC Genome Browser and GeneRunner Software. Following settings were used to design primers:

Primer Tm:	65-68 °C
Primer length:	18-28 bp
Amplicon size:	80-150 bp

4.3.4.2 Standard PCR for cloning or sequencing of gDNA

PCRs were generally performed in "thick" PCR tubes with a reaction volume of 20-100 µl in a MJ research PTC 200 thermocycler (Biozym). The "calculated temperature" feature

Methods

was used to decrease temperature hold times and additionally the lid was heated to 105°C to prevent vaporisation. The nucleotide sequences of the utilized primers are given in section 3.6. The primer annealing temperatures varied between 57 and 65°C. For a typical reaction the Phusion™ Hot Start High-Fidelity DNA Polymerase (Finnzymes; Espoo, Finnland) was used with the following basic reaction conditions:

Component	Volume	Final concentration
H_2O	Add 50 µl	
5×Phusion HF buffer	10 µl	1×
10 mM dNTPs	1 µl	200 µM each
Primer S	1 µl	0.2 µM
Primer AS	1 µl	0.2 µM
Template DNA	X µl	
Phusion Polymerase (2 U/ml)	0.5 µl	0.02 U/µl

General parameter settings for analytical PCR are summarized in Table 4-5 Reaction parameter for analytical PCR.

Table 4-5 Reaction parameter for analytical PCR

PCR step		Cycling parameter
Initial melting		95°C, 2 min
20 - 35 cycles	Melting	95°C, 15 s
	Annealing	65°C, 15 s
	Extension	72°C, 60 s
Final extension		72°C, 5 -7 min
Cooling		15°C, forever

4.3.4.3 Real-time PCR

Quantitative real-time PCR (qPCR) enables both detection and quantification (as absolute number of copies or relative amount when normalized to DNA input or additional normalizing genes) of a specific sequence in a DNA sample. The procedure follows the general principle of polymerase chain reaction; its key feature is that the amplified DNA is quantified as it accumulates in the reaction in *real-time* after each amplification cycle. On the one hand, this method was used to quantify cDNA after reverse transcription (see section 4.3.5.3), on the other hand to quantitatively analyze genomic DNA after fragmentation by methyl-CpG immunoprecipitation (MCIp, see section 4.4.4) or after chromatin immunoprecipitation. PCR reactions were performed using the Quantifast SYBR Green Kit from Qiagen either in glass capillaries using the LightCycler system from Roche (total volume: 20 µl) or in 96-well format adapted to the Eppendorf Realplex Mastercycler EpGradient S (Eppendorf, Hamburg, Germany). The relative amount of amplified DNA is measured through the emission of light by the SYBR green dye when it is intercalated in double-stranded DNA.

Reaction setup:
    5 µl SYBR Green mix (2×)
    2 µl ddH$_2$O
    0.5 µl sense primer
    0.5 µl antisense primer
    2 µl DNA

Table 4-6 Reaction parameter for real-time PCR

PCR step		Cycling parameter
Initial melting		95 °C, 5 min
45 cycles	Melting	95 °C, 8 s
	Annealing & extension	60 °C, 20 s
Final cycle	Melting	95 °C, 15 s
	Annealing & extension	60 °C, 15 s
Melting curve		10 – 20 min
		95 °C, 15 s

To calculate amplification efficiency, a standard curve was generated for each primer pair by amplifying four dilutions (1:10, 1:50, 1:100, 1:1000). Realplex software calculated automatically DNA amounts based on the generated *slope* and *intercept*. Specific amplification was controlled by melting-curve analysis and data was imported and processed in Microsoft Excel 2007. All samples were measured in duplicates and normalized to the ß-actin or the HPRT housekeeper when analyzing mRNA expression. Duplicates of ChIP samples were normalized to the input or a not affected upstream control region.

Methods

4.3.4.4 MassARRAY quantitative gene expression (QGE) analysis

4.3.4.4.1 Principle

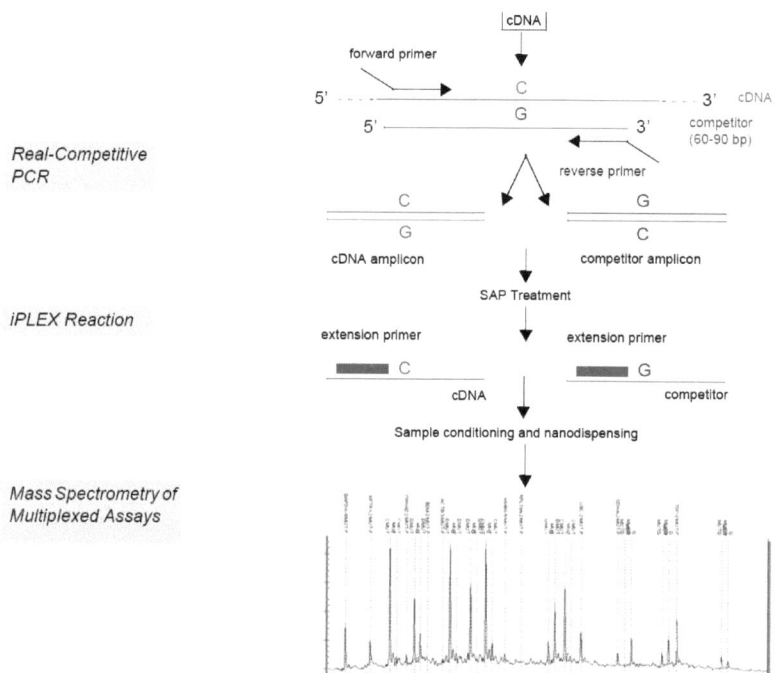

Figure 4-1 Schematic outline of the MassARRAY QGE process
cDNA or MCIp-enriched DNA is spiked with a synthetic DNA molecule (competitor), which matches the sequence of the targeted cDNA region in all positions except a single base and serves as internal standard. The cDNA/competitor is PCR-amplified and subjected to a SAP treatment. After inactivation of the SAP, a primer extension cocktail is added. The PCR products from the competitor and the cDNA now serve as templates for the iPLEX reactions. After primer extension, the products are desalted using clean resin and then dispensed on a SpectroCHIP for subsequent MALDI-TOF MS (www.sequenom.com).

The MassARRAY QGE method combines real-competitive PCR (rcPCR) with the iPLEX primer extension reaction, followed by matrix-assisted laser desorption/ionization time-of-flight mass spectrometry (MALDI-TOF MS). cDNA or MCIp-enriched DNA is spiked with a competitor, which matches the sequence of the targeted cDNA region in all positions except a single base and serves as an internal standard. DNA and competitor are PCR-amplified and then SAP (shrimp alkaline phosphatase) -treated to dephosphorylate remaining nucleotides. After SAP inactivation, a primer extension cocktail is added. The PCR products from the competitor and the cDNA now serve as templates for the iPLEX reactions. After primer extension, the products are desalted using clean resin and then dispended on a SpectroCHIP for subsequent MALDI-TOF MS. During mass spectrometric analysis, the peak areas of the distinct mass signals for the competitor and DNA extension products are resolved and peak area ratios are calculated. The QGE Analyzer software plots cDNA frequency versus competitor concentration for each assay and sample. DNA concentrations (expressed as LOGEC50 or EC50) are automatically calculated via non-linear regression analysis and represent the competitor concentration at which the allele frequencies of cDNA and competitor are equal (0.50:0.50). A workflow for conducting MassARRAY QGE experiments is shown in Figure 4-1. Detailed description of the method is given in the MassARRAY QGE-iPLEX Application guide (www.sequenom.com).

4.3.4.4.2 Protocol

Primer and competitive template designs were created using the MassARRAY QGE Assay Design software v1.0 (Sequenom, San Diego, CA). Preparing of the competitor plates, PCR, SAP addition, iPLEX reaction, desalting of the iPLEX reaction and MALDI-TOF analysis were performed as described in the Sequenom protocols. Raw data were then processed using the MassARRAY QGE Analyzer software v3.4.

4.3.4.5 Nested PCR for quantitative methylation analysis

Methylation analysis of specific DNA fragments was performed using a nested PCR after bisulfite treatment of genomic DNA (see section 4.4.3). 10 µl of bisulfite-treated DNA were used for the first nested PCR reaction, generated with an outer primer pair. Afterwards the PCR products are used as a template for a second PCR using a different set of primers inside of the first PCR product (inner primer pair). The reaction was performed as follows:

Components	1st PCR	2nd PCR
Bisulfite DNA	10 µl	
DNA from 1st PCR		0.5 µl
10×Taq-Buffer	5 µl	5 µl
dNTPs (10 mM each)	1 µl	1 µl
Out S (10 µM)	2 µl	
Out AS (10 µM)	2 µl	
In S (10 µM)		2 µl

		2 µl
In AS (10 µM)		2 µl
Taq	0.5 µl	
FastStart-Taq		0.5 µl
H$_2$O	Ad 50 µl	Ad 50 µl

Table 4-7 Reaction parameter for nested PCR

step	1st PCR			2nd PCR		
	Temp	Time	cycle	Temp	Time	cycle
Initial denaturation	93 °C	5 s	1	94 °C	3 min	1
Denaturation	93 °C	15 s		94 °C	15 s	
Annealing	55 °C	15 s	30	55 °C	15 s	30-35
Extension	72 °C	70 s		72 °C	80 s	
Final extension	72 °C	5 min	1	72 °C	5 min	1

After the second amplification reaction, products were cloned into a TOPO vector and transformed in TOP10 cells (one shot chemical transformation) (see section 4.3.1.2). After preparation of plasmid DNAs, samples were then sent to Entelechon or Geneart for sequencing.

4.3.5 Preparation and analysis of RNA

4.3.5.1 Isolation of total RNA

Total RNA was isolated using the Qiagen RNeasy Midi, Mini or Micro Kit according to the available number of cells. RNA concentration was then determined with the NanoDrop spectrophotometer and quality was assessed by agarose gel electrophoresis or using the Agilent Bioanalyzer according to the manufacturer's instructions.

4.3.5.2 Formaldehyde agarose gel

The agarose was dissolved in MOPS/H$_2$O$_{DEPC}$ by heating in a microwave oven and cooled to 60 °C. Formaldehyde was added while stirring the solution under a fume hood and the gel was cast, mounted in an electrophoresis tank and overlaid with 1×MOPS as electrophoresis buffer. RNA samples were heated to 37 °C for 30 min to control RNase contamination and placed on ice afterwards. Samples were subsequently diluted with four volumes RNA loading buffer (1:4), denatured for 20 min at 65 °C and briefly incubated on ice. Following centrifugation, the samples were loaded into the gel slots. Gels were run at 40-60 V.

Required buffers

MOPS (20×)	42 g	(0.4 M)	MOPS/NaOH, pH 7.0
	4.1 g	(100 mM)	NaOAc
	3.7 g	(20 mM)	EDTA
	Add H_2O_{DEPC} to 500 ml, store in the dark		
RNA loading buffer	10 ml	(50%)	Formamide, deionized
	3.5 ml	(2.2 M)	Formaldehyde (37%)
	1 ml	(1×)	MOPS (20×)
	0.8 ml	(0.04%)	Bromophenol blue (1% in H_2O)
	0.2 g	(1%)	Ficoll 400, Pharmacia (dissolve in 2 ml H_2O)
	Add H_2O_{DEPC} to 20 ml, store in 1 ml aliquots at -20 °C		

Add 5 μl/ml ethidium bromide (10 mg/ml) before use

4.3.5.3 Reverse transcription PCR (RT-PCR)

To quantify mRNA transcripts of genes, total RNA was reverse transcribed using the MMLV reverse transcriptase (Promega, Germany) combined with random decamers (Ambion, Germany) in a total reaction volume of 20 μl.

Reaction setup:
- 1 μg Total RNA
- 1 μl Random decamers
- 1 μl dNTPs (10 pmol/ml)
- Add H_2O_{USB}

Incubate for 5 min at 65 °C, cool on ice and centrifuge

- 4 μl M-MLV Buffer (5×)

Mix and incubate for 2 min at 42 °C

- 1 μl M-MLV Reverse transcriptase

Incubate for 50 min at 42 °C followed by 15 min at 70 °C

The resulting cDNA was then quantified with specific primers by real-time PCR (see section 4.3.4.3).

Methods

4.3.5.4 Whole genome gene expression

4.3.5.4.1 4 × 44K Agilent whole human genome expression array

Labeling, hybridization and scanning of high quality RNA were performed using the Agilent Gene Expression system according to the manufacturer's instructions. In brief, 200 ng to 1000 ng high-quality RNA were amplified and Cyanine 3-CTP labeled with the One colour Low RNA Input Linear Amplification Kit from Agilent. Labeling efficiency was controlled using the NanoDrop spectrophotometer and 1.65 µg labeled cRNA was fragmented and hybridized on the whole human genome expression array (4 × 44K, Agilent). After 17 hours of hybridization at 65 °C, the microarrays were washed and subsequently scanned with an Agilent scanner. Data were then extracted with Feature Extraction 9.5.1 software (GE1 v5_95_Feb07 protocol, Agilent) and finally analyzed using GeneSpring G 7.3.1 software (Agilent). To validate microarray data, several genes were selected and verified by RT-PCR followed by qPCR (see sections 4.3.5.3 and 4.3.4.3).

4.3.5.4.2 Affymetrix microarray analysis

RNA from KG-1, U937, and THP-1 cells as well as from freshly isolated human blood monocytes of healthy donors were alternatively (formerly) analyzed using Affymetrix HG-U133_Plus_2 arrays. Hybridization, cRNA labeling and data handling was done by the KFB (Regensburg).

4.3.6 ChIP-on-chip

Chromatin immunoprecipitation (ChIP) is a method to investigate interactions between proteins and DNA *in vivo*. Therefore DNA is covalently bound to proteins with formaldehyde, fragmented by sonication and precipitated with suitable antibodies. Hereafter, the covalent cross-links are broken up to free the precipitated DNA. The quality of each ChIP was controlled at known target sites by qPCR. If the ChIP was successful the fragments were amplified with a ligation-mediated PCR (LM-PCR, see section 4.3.6.2), fluorescence-labeled and hybridized on a microarray (ChIP-on-chip, microarray handling see section 4.4.5) against a fractional amount of the input to correct background noise. This approach allows the identification of binding sites of DNA-binding proteins in all areas covered on the microarray platform.

4.3.6.1 Chromatin immunoprecipitation (ChIP)

Preparation of cross-linked chromatin was performed as described previously with some modifications. In summary, 2 million cells were used per immunoprecipitation. The cells were treated with 1% formaldehyde solution for 10 min at room temperature and quenched by 0.125 M glycine. After washing with PBS including 1 mM PMSF, 2×10^6 cells were resuspended in 50 µl lysis buffer 1A (L1A: 10 mM, HEPES/KOH, pH 7.9, 85 mM KCl, 1 mM EDTA, pH 8.0) and lysed by adding 50 µl lysis buffer 1B (L1A + 1% Nonidet P-40) for 10 min on ice. Note that lysis buffers were supplemented with phosphatase inhibitors (50 mM ß-glycerophosphate and 1 mM Na_2OV_4) when phosphorylated proteins had to be precipitated. The lysate was centrifuged (700×g, 5 min), the supernatant discarded and the nuclei were resuspended in 400 µl L2. If more cells were available, up to 20×10^6 cells were treated with the same procedure to concentrate the chromatin. Cross-linked chromatin was sheared to an average DNA fragment size around 400 – 600 bp using a Branson Sonifier 250 (Danbury, CT). The sonicated lysate was cleared by centrifugation (13,000 rpm, 5 min, 4°C) and the supernatant was transferred into a new 1.5 ml tube. To monitor successful fragmentation of the DNA an aliquot was taken for agarose gel analysis (which was incubated overnight with 200 mM NaCl at 65 °C to reverse the formaldehyde cross-links and purified with the QIAquick PCR purification kit (Qiagen)) and a 5 % volume aliquot of the lysate was kept as the input. To pre-clear the lysate 50 µl/precipitation sepharose CL-4B beads were washed twice with TE pH 8.0, filled up with dilution buffer to the previous volume and incubated with 25 µl 20% BSA and 4 µl glycogen per ml CL4Beads on a rotator for a minimum of 2 hours at room temperature. The lysate was diluted 1:1.5 with DB and 50 µl of the CL-4B beads/ IP were added, rotating for 2 hours at 4 °C.

Following this the pre-cleared lysate was recovered by centrifugation (13,000 rpm, 5 min, 4°C) and 200 µl supernatant for each IP was transferred in a new 0.5 ml PCR tube. Antibodies were added (2-5 µg each, depending on the used cell numbers) and incubated on a rotator at 4 °C overnight.

To bind the antibody complexes to beads, 55 µl nProtein A sepharose beads per IP were washed twice with TE pH 8.0, filled up to the previous volume with DB and blocked with 0.4 µl glycogen and 2.5 µl BSA (20%) per 100 µl beads overnight on a rotator at 4°C. Then 50 µl of the blocked beads were added to the antibody complexes, rotated at 4°C for 2 hours, centrifuged (4,000 rpm, 5 min, 4 °C) and the supernatant was discarded. The beads were transferred on Millipore Ultrafree-MC columns and washed twice with WBI, WBII, WBIII and three times with TE pH 8.0, shaking the beads for 5 minutes in between. The

Methods

DNA was eluted in two steps by adding 100 µl EB each, incubating for 20 minutes and 10 minutes respectively, shaking up the beads every 5 minutes. 200 µl EB was added to the input as well, and all samples were incubated overnight at 65 °C with added Proteinase K (0.5 µg/µl final concentration, Roche) to reverse the cross-links. RNase (0.33 µg/µl, Qiagen) digestion for 2 hours at 37 °C degraded RNA that could interfere with downstream applications. Finally, the samples were purified with the QIAquick PCR purification kit following the manufacturer's instructions with small variations: binding buffer PB was incubated for 30 minutes, binding DNA to the column by centrifugation was carried out at 10,000 rpm and elution was done with 100 µl pre-warmed elution buffer EB.

Required buffers and solutions:

Glycine	9.85g (2.625 M)	Glycine	
	Ad 50 ml with ddH$_2$O		
Cell Buffer Mix	20 µl (10 mM)	HEPES / KOH (1 M), pH 7.9	
	57 µl (85 mM)	KCl (3 M)	
	4 µl (1 mM)	EDTA (0.5 M, pH 8.0)	
	Ad 1.98 ml with ddH$_2$O		
	Add just prior to use:		
	20 µl (1 mM)	PMSF (100 mM in Iso-prop, nostalgia)	
	2 µl (1 µg/ml)	Pepstatin (1 µg/µl)	
	2 µl (2 µg/ml)	Aprotinin (2 µg/µl)	
Nuclear Lysis Buffer (L2)	100 µl (50 mM)	Tris/HCl (1 M), pH 7.4 @ 20 °C	
	100 µl (1%)	SDS (20%)	
	33.3 µl (0.5%)	Empigen BB (30%)	
	40 µl (10 mM)	EDTA (0.5 M), pH 8.0	
	Ad 1.98 ml with ddH$_2$O		
	Add just prior to use:		
	20 µl (1 mM)	PMSF (100 mM in Iso-prop, nostalgia)	
	2 µl (1 µg/ml)	Pepstatin (1 µg/µl)	
	2 µl (2 µg/ml)	Aprotinin (2 µg/µl)	
Dilution Buffer (DB)	50 µl (20 mM)	Tris/HCl (1 M), pH 7.4 @20 °C	
	50 µl (100 mM)	NaCl (5 M)	
	10 µl (2 mM)	EDTA (0.5 M, pH 8.0)	
	125 µl (0.5%)	Triton X-100 (10%)	
	Ad 2.47 ml with ddH$_2$O		
	Add just prior to use:		
	25 µl (1 mM)	PMSF (100 mM in Iso-prop, nostalgia)	
	2.5 µl (1 µg/ml)	Pepstatin (1 µg/µl)	
	2.5 µl (2 µg/ml)	Aprotinin (2 µg/µl)	

Wash Buffer I (WB I)	200 µl (20 mM) 300 µl (150 mM) 50 µl (0.1%) 1 ml (1%) 40 µl (2 mM) To 10 ml with ddH₂O	Tris/HCl (1 M), pH 7.4 @ 20 °C NaCl (5 M) SDS (20%) Triton X-100 (10%) EDTA (0.5 M, pH 8.0)
Wash Buffer II (WB II)	200 µl (20 mM) 1 ml (500 mM) 40 µl 1 ml (1%) 40 µl (2 mM) To 10 ml with ddH₂O	Tris/HCl (1 M), pH 7.4 @ 20 °C NaCl (5 M) SDS (20%) Triton X-100 (10%) EDTA (0.5 M, pH 8.0)
Wash Buffer III (WB III)	100 µl (10 mM) 250 µl (250 mM) 1 ml (1%) 1 ml (1%) 20 µl (1 mM) To 10 ml with ddH₂O	Tris/HCl (1 M), pH 7.4 @ 20 °C LiCl (10 M) hard to dissolve, try 2.5 M NP-40 (10%) Deoxycholate (10%) EDTA (0.5 M, pH 8.0)
Elution Buffer (EB)	450 µl (0.1 M) 225 µl (1%) To 4.5 ml with ddH₂O	NaHCO₃ (1M) SDS (10%)

4.3.6.2 LM-PCR

Ligation mediated PCR (LM-PCR) was used to amplify the chromatin immunoprecipitated DNA. Adaptors are ligated to all fragments in the precipitation, and primers specific for these adaptors are used to amplify all fragments independent of their sequences. All reagents were purchased from New England Biolabs (NEB) unless otherwise mentioned.

To prepare the 60 mM linker, 10 ml Tris-HCl (1 M) pH 7.9, 15 µl oligo JW102_sticky and 15 µl oligo JW103 (160 µM each, Metabion) were mixed and incubated in a thermocycler with the following program:

Step 1	95 °C	5 min
Step 2	70 °C	1 min
Step 3	Ramp down to 4 °C (0.4 °C /min)	
Step 4	4 °C	HOLD

To start, the overhangs were converted into phosphorylated blunt ends, using T4 DNA polymerase, *E.coli* DNA Pol I large fragment (Klenow polymerase), and T4 polynucleotide kinase (PNK). The 3' to 5' exonuclease activity removes 3' overhangs, the polymerase activity fills in the 5' overhangs and the PKN adds the phosphate group to the 3' end.

ChIP enriched DNA (about 10 ng) was brought to a volume of 40 µl with ddH$_2$O. Then 10 µl of the reaction mix was added:

T4 DNA Ligase buffer with 10 mM ATP	(5 µl)
dNTP mix	(2 µl)
T4 DNA polymerase	(1 µl)
Klenow DNA polymerase diluted with water to 1 U/µl	(1 µl)
T4 PNK	(1 µl)

The mixture was incubated in a thermocycler for 30 minutes at 20 °C, then purified with the QIAquick PCR purification kit (Qiagen) and finally eluted in 34 µl elution buffer (EB). The eluate was then incubated with 1 µl of Klenow fragment (3' to 5' exo minus), 5 µl NEB buffer II and 10 µl dATP (1 mM) for 30 minutes at 37 °C, followed by clean-up with the MinElute kit (Qiagen), eluting in 10 µl EB. In this process an adenine overhang was added to the DNA fragments' 3' ends to facilitate the ligation with the adapters, which have a single "T" base overhang at their 3' site (see oligo JW102_sticky). DNA Quick-Ligase buffer 2 (15 µl), linker 60 µM preparation (1 µl) and DNA Quick-Ligase (4 µl) were mixed with the DNA sample and incubated for 15 minutes at room temperature. The reaction was cleaned up with the QIAquick PCR purification kit (Qiagen) and eluted in 25 µl EB. For large-scale amplification of IP samples two buffer mixes were prepared:

Mix A:

Stock	1× Mix	Final Concentration
5×Phusion polymerase buffer	8.00 µl	1×
dNTP mix (10 mM each)	1.25 µl	250 µM
Oligo JW102_sticky (160 µM)	0.31 µl	1 µM
Betaine	5.44 µl	1.5 M
Total	15 µl	

Mix B:

Stock	1× Mix	Final Concentration
5×Phusion polymerase buffer	2.00 µl	1×
Phusion Polymerase(2 U/µl)	0.50 µl	1 U
Betaine	7.50 µl	1.5 M
Total	10.00 µl	

15 µl of Mix A was added to the sample and on a thermocycler the following program was started:

Table 4-8 Reaction parameter for 1st LMPCR

PCR step		Cycling parameter
Initial heating		55 °C, 4 min
End-filling		72 °C, 30 s
Initial melting		98 °C, 30 s
15 cycles	Melting	98 °C, 10 s
	Annealing	68 °C, 30 s
	Elongation	72 °C, 30 s
Final elongation		72 °C, 5 min
Cooling		4 °C, forever

Midway through step 1 (initial heating) the program was paused and 10 µl Mix B was added to "hot start" the reaction. The PCR product was diluted with 475 µl ddH$_2$O, and 5 µl were used for a second expansion using the following mixture:

Stock	1× Mix	Final concentration
5× Phusion polymerase buffer	10.00 µl	1×
dNTP mix (10 mM each)	1.25 µl	250 µM
Oligo JW102_sticky (160 µM)	0.31 µl	1 µM
ddH$_2$O	19.94 µl	
Phusion Polymerase (2 U/µl) HOT START	0.50 µl	1 U
PCR dilution (first amplification)	5 µl	
Betaine	13 µl	1,5 M
Total volume	50 µl	

Methods

The PCR program for the second expansion was:

Table 4-9 Reaction parameter for 2nd LMPCR

PCR step		Cycling parameter
Initial melting		98°C, 30 s
25 cycles	Melting	98°C, 10 s
	Annealing	68°C, 30 s
	Elongation	72°C, 30 s
Final elongation		72°C, 5 min
Cooling		4°C, forever

The product was cleaned up with the QIAquick PCR purification kit (Qiagen) and eluted in 50 µl EB. DNA concentration was measured with the NanoDrop instrument (Peqlab).

4.3.6.3 Labeling and hybridization

Amplified ChIP material and genomic input were labeled with Alexa Fluor 5-dCTP and Alexa Fluor 3-dCTP, respectively. Comparative ChIP-versus-input hybridizations on CpG island oligonucleotide microarrays (Agilent) were performed using the recommended, stringent protocol (see section 4.4.5.2).

4.4 Analysis of DNA methylation

4.4.1 *In vitro* methylation of DNA

10-20 µg plasmid or genomic DNA were incubated with 2.5 U/µg *Sss* I methylase in the presence of 160 µM S-adenosylmethionine (SAM; methyl group donor) for 4 hours at 37°C. After 2 hours the reaction was supplied with another 160 µM SAM. Simultaneously, control reactions were treated as above but without addition of SAM and methylating enzymes. After the methylation reaction, DNA was purified using the NucleoSpin® Plasmid Quick Pure Kit from Macherey-Nagel or by phenol-chloroform extraction followed by ethanol precipitation and finally quantified using a NanoDrop spectrophotometer. Completeness of methylation was controlled by digesting both methylated and unmethylated DNA using the methylation-sensitive restriction enzymes *Hha* I and *Hpa* II.

4.4.2 Generation of an *in vitro* partially methylated gene locus

A fragment of the CpG island promoter of *CPM* was subcloned into the CpG-free plasmid pCpG-mcs (Invivogen). The plasmid was linearized with *Ase* I to generate a fragment of the CpG island promoter flanked with CpG-less sequences on either side. The DNA fragment was then treated with *Sss* I (New England Biolabs) and decreasing amounts of the methyl donor S-adenosylmethionine (160 µM, 40 µM, 10 µM, 2.5 µM, 0.7 µM). Samples were combined to obtain a mixture of DNA fragments with varying density of CpG methylation.

4.4.3 Bisulfite sequencing

A common method for analyzing cytosine methylation is bisulfite conversion of DNA followed by sequencing. Cytosine-derivates undergo reversible reactions with bisulfite yielding a 5,6-Dihydro-6-sulfonate, which deaminates spontaneously. After that the sulfate is eliminated under alkaline conditions, leaving uracil.
5'-methylcytosine is not affected by this reaction. Modification of gDNA with sodium bisulfite, leading to conversion of unmethylated cytosine residues into uracil while not affecting 5-methylcytosine (Frommer et al., 1992a), was performed using the Qiagen EpiTect Bisulfite Kit as recommended by the manufacturer. 10 µl of bisulfite-treated DNA were used for the nested PCR reaction (see section 4.3.4.5).

4.4.4 Methyl-CpG immunoprecipitation (MCIp)

The MCIp is a method that allows the rapid and sensitive screening of DNA methylation. The application consists of the binding of methylated DNA fragments to the bivalent, antibody-like fusion protein MBD-Fc (a methyl binding domain fused to an Fc-tail) in an immunoprecipitation-like approach. The affinity to DNA is increased with the density of methylated CpGs and lowered with higher salt concentrations in the buffer. Washing with buffers containing increasing NaCl concentrations and collection of according flow-throughs leads to the fragmentation of DNA depending on the methylation status of CpG dinucleotides. Enriched methylated DNA fragments can be efficiently detected on single gene level or on a genome-wide level. The recombinant MBD-Fc protein was produced as previously described (Gebhard, 2005; Gebhard et al., 2006b; Gebhard et al., 2006a) (see

Methods

also section 4.1.2) and MCIp was performed with slight modifications. A schematic representation is given in Figure 5-1 and Figure 5-8.

Required buffers and solutions:

TME (10×)	200 mM	Tris-HCl (1 M) pH 8.0
	20 mM	$MgCl_2$ (1 M)
	5 mM	EDTA (500 mM)
Buffer A	1×	TME (10×)
(300 mM NaCl)	300 mM	NaCl (5 M)
	0.1%	NP-40 (10%)
Buffer X	1×	TME (10×)
(300 mM NaCl)	300 mM	NaCl (5 M)
	0.1%	NP-40 (10%)
Buffer B-G	1×	TME (10×)
	0.1%	NP-40 (10%)
	350 (B), 400 (C), 450 (D), 500 (E), 600 (F), 1000 mM (G)	

4.4.4.1 DNA fragmentation

Genomic DNA was either restriction digested with *Mse* I or sonicated to a mean fragment size of 350-400 bp. Before sonication gDNA was initially sheared using a 20 gauge needle attached to a 2 ml syringe (BD) before quantification using the NanoDrop ND 1000 spectrophotometer (Peqlab). Sonication was carried out with the Branson Sonifier 250 (Danbury) using the settings shown below. After sonication the sample was immediately cooled on ice. The fragment range was controlled using agarose gel electrophoresis.

For 5 µg DNA in 500 µl TE	Duty cycle	30%
	Output	3
	Sonication time	60 s
For 2.5 µg DNA in 500 µl TE	Duty cycle	80%
	Output	0.5
	Sonication time	2×30 s

4.4.4.2 Binding MBD2-Fc to beads

For small-scale reactions, typically 13-18 µg purified MBD-Fc protein per 40 µl nProtein A-Sepharose 4 Fast Flow beads (Amersham Biosciences) were rotated in 2 ml TBS overnight at 4°C in order to bind the Fc-part of the protein to the beads. For large-scale reactions 60-80 µg MBD-Fc protein was bound to 150-200 µl nProtein A–Sepharose 4 Fast Flow beads. On the next day, the MBD2–Fc-bead complexes (40 µl/assay) were transferred and dispersed equally into 0.5 ml (for large-scale reaction: 2 ml) Ultrafree-MC centrifugal filter devices (Millipore) and spin-washed twice with buffer A.

4.4.4.3 Enrichment of highly methylated DNA

For small-scale reactions digested or sonicated DNA (150-300 ng) was added to the washed MBD2–Fc beads in 350 µl buffer and rotated for 3 h at 4 °C. Beads were centrifuged to recover unbound DNA fragments (300 mM fraction) and subsequently washed twice with 200 µl and 150 µl of buffers containing increasing NaCl concentrations (350-1000 mM, see buffers B-G). The flow-through of each washing step was collected in separate tubes and desalted using a QIAquick PCR Purification kit (Qiagen). In parallel, 150-300 ng fragmented input DNA was resuspended in 350 µl buffer and desalted using a QIAquick PCR Purification kit (Qiagen) as a control.

To generate DNA fragments for direct labeling for microarray hybridization this MCIp protocol was scaled up. For large-scale reactions, for each sample, 60-80 µg purified MBD2–Fc protein was added to 150-200 µl Protein A–Sepharose beads (Amersham Biosciences) in 15 ml TBS and rotated overnight at 4°C. For the precipitation, 2 ml Ultrafree-MC centrifugal filter devices (Millipore) were used and 2 or 4 µg of sonicated DNA (in large scale reactions no digested DNA was used). The flow-throughs were collected and desalted using a QIAquick PCR Purification kit (Qiagen) or the MinElute Kit. In parallel, 1/10 of the DNA used for precipitation was also desalted using the respective kit and used as input DNA. The separation of CpG methylation densities of individual MCIp fractions was controlled by qPCR using primers covering the imprinted *SNRPN* gene and a region without any CpGs (Empty 6.2), respectively. For the microarray approach a threshold was defined and flow-throughs were combined to a hypermethylated fraction for subsequent labeling and microarray analysis.

4.4.5 DNA Microarray handling and analysis

4.4.5.1 Human CpG 12K arrays

To generate fluorescently labeled DNA for CpG island microarray hybridization, Mse I-compatible uni-directional LMPCR linker (LMPCR_S-L 5'-GCG GTG ACC CGG GAG ATC TCT TAA G-3' and LMPCR_AS-L: 5'-TAC TTA AGA GAT C-3', 20 µM) were ligated to the MCIp-eluted DNA and in a separate reaction to an equal amount of input DNA (0.5 µl linker /ng DNA) in 60 µl reactions using 1,200 U T4 Ligase (NEB) at 16°C O/N. Linker-ligated DNA was desalted using QIAquick PCR Purification kit (Qiagen). Amplification of linker-ligated DNA preparations was performed using LMPCR primer (5'-GTG ACC CGG GAG ATC TCT TAA G-3') and Taq polymerase (Roche) in the presence of 1.3 M betaine. Amplicons were desalted using QIAquick PCR Purification kit (Qiagen) and quantified (PicoGreen dsDNA Quantitation Reagent, Molecular Probes). Labeling and hybridization of MCIp amplicons were done by the KFB (Regensburg) according to the protocol provided by the CpG island microarray manufacturer (Microarray Centre UHN, Toronto, Canada) with modifications. Briefly, four microgram of normal and tumor MCIp-amplicons were directly labeled with Cy5- and Cy3-dCTP, respectively, using the BioPrime® Array CGH Genomic Labeling System (Invitrogen). Ten microgram of each fluorescently labeled and purified DNA amplicon in 300 µl DIG Easy Hyp Solution (Roche) supplemented with 25 µg Cot-1 DNA (Invitrogen) and 30 µg Yeast tRNA were hybridized to Human CpG 12K Arrays (HCGI12K, Microarray Centre, UHN, Toronto, Canada) in 6021 mm Gene Frames (ABgene) at 37°C for O/N. Slides were washed three times in 1×SSC, 0.1% SDS at 50°C for 10 min. After two more rinses with 0.1×SSC, slides were dried and scanned using the Affymetrix 428 Scanner. Images were analyzed using the ImaGene 5.6 and Gene Sight Lite software (BioDiscovery, Inc., EL Segundo, CA). Locally weighted scatter plot smoothing normalization was used to normalize Cy3 and Cy5 signals. Clones that produced reproducible differential signals on the CpG island microarray were sequenced by the University Health Network Microarray Centre.

4.4.5.2 Human 244K Agilent CpG island microarrays

4.4.5.2.1 Labeling reaction

Enriched methylated DNA fragments of the high salt MCIp fractions were labeled with Alexa Fluor 5-dCTP (cancer cells) and Alexa Fluor 3-dCTP (normal cells) using the

BioPrime Total Genomic Labeling System (Invitrogen) as indicated by the manufacturer. Amplified ChIP material was labeled with Alexa Fluor 5-dCTP and the genomic input with Alexa Fluor 3-dCTP. Labeling efficiency was controlled using the NanoDrop Nd-1000 spectrophotometer (PeqLab, Erlangen, Germany).

4.4.5.2.2 Microarray hybridization

The differently labeled DNA fragments or pools of two samples were combined and supplemented with human Cot-1 DNA, Agilent blocking agent (10-fold) (Agilent Technologies, Böblingen, Germany), Agilent hybridization buffer (2-fold) as supplied in the Agilent oligo aCGH Hybridization Kit. For more stringent hybridization conditions deionized formamide was additionally added in order to prevent cross-hybridization of GC-rich DNA sequences.

Component	1x Mix	Final concentration
DNA samples combined in 80 µl TE	80 µl	
Cot-1 DNA (1.0 mg/ml)	50 µl	0.1 mg/ml
Agilent Blocking agent (10×)	52 µl	1×
Deionized formamide	78 µl	15%
Agilent Hybridization Buffer (2×)	250 µl	1×

The sample was heated to 95°C for 3 min, mixed, and subsequently incubated at 37°C for 30 min and spun down afterwards for 1 min. Hybridization on microarray slides (Agilent) was then carried out at 67°C for 40 h using an Agilent SureHyb chamber and an Agilent hybridization oven. Slides were washed in Wash I (6×SSPE, 0.005% N-lauroylsarcosine) at room temperature for 5 min and in Wash II (0.06×SSPE; prewarmed to 37°C for stringent protocol) for additional 5 min. Afterwards slides were dried and incubated using acetonitrile for 30 s. Images were scanned immediately and analyzed using a DNA microarray scanner (Agilent). Microarray images were processed using Feature Extraction Software 9.5.1 (Agilent) using the standard CGH protocol for samples from MCIp. Processed data were imported into Microsoft Office Excel for further analysis. Graphical presentations of datasets were obtained using Spotfire Decision Site Software 7.0 (Spotfire).

4.4.6 Quantitative DNA methylation analysis using the MassARRAY system (SEQUENOM)

4.4.6.1 Principle

Quantitative assessment of DNA methylation in target genomic regions was performed using the Sequenom MassCLEAVE™ assay. DNA samples for analysis are initially bisulfite-treated, resulting in the conversion of unmethylated cytosines to uracil, whereas methylated cytosines remain unchanged. This conversion reaction allows for accurate discrimination between methylated and unmethylated cytosines at CpG dinucleotides. Following bisulfite treatment, genomic DNA consists of two non-complementary single-stranded DNA populations. Subsequently, PCR primer pairs for a region of interest are designed to amplify both the forward and reverse strand of double-stranded genomic DNA. A T7 polymerase promoter tag is added to the 5' end of the reverse primer to facilitate *in vitro* transcription and a 10-mer tag is added to the 5' end of the forward PCR primer to minimize melting temperature differences between both primers during PCR cycling. Following PCR, unincorporated dNTPs are dephosphorylated by treatment with SAP. Reverse transcription is performed using a chemically modified T7 RNA polymerase which utilizes a mixture of ribonucleotides and deoxyribonucleotides when synthesizing the RNA strand. In parallel with the reverse transcription the cleave reaction is achieved using the pyrimidine specific Ribonuclease A (RNaseA) enzyme which cleaves at pyrimidines (C and T) only on the newly synthesized transcript. By incorporating a non-cleavable dCTP (deoxyribonucleotide) into the transcript, RNaseA is unable to cleave at C and can only cleave at T (T specific cleavage) yielding a population of single-stranded cleavage fragments (Figure 4-2). A methylated cytosine is represented by a G nucleotide in the cleavage fragment, whereas an unmethylated cytosine is represented by an A nucleotide. The mass difference of 16 Da between G (329 Da) and A (313 Da) is easily detected by MALDI-TOF MS. Depending on the number of methylated CpG sites within a cleavage fragment, the difference in mass will increase in 16 Da units.

As already mentioned, in the following procedure this methylation specific difference is not used for sequencing (see section 4.4.3) but for generating methylation depending mass differences to be analyzed by mass spectrometry. A detailed description of the method is given in Ehrich et al., 2005 and in the EpiTYPER User Guide (www.sequenom.com).

Methods

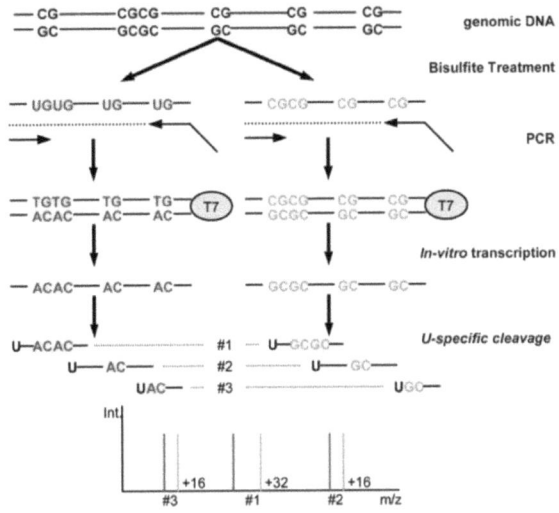

Figure 4-2 Schematic outline of the EpiTYPER process
Genomic DNA is treated with bisulfite and amplified using specific primers with one primer tagged with a T7 promoter sequence. PCR products are subsequently transcribed into RNA, followed by RNase cleavage after every uracil residue. Cleavage products are then analyzed by MALDI-TOF MS. In the example shown here, PCR products are transcribed from the reverse strand. In the unmethylated template (illustrated in red) cytosine residues are deaminated into uracil and therefore appear as adenosine residues after PCR. Cytosine residues of a methylated template (indicated in yellow) are not affected and remain cytosines. The conversion of guanine to adenine yields 16 Da mass shifts. Cleavage product 1 comprises 2 CpGs and the mass difference constitutes 32 Da if both CpGs are either methylated or unmethylated. Cleavage products 2 and 3 each contain only one CpG site that is differentially methylated and therefore yield a 16 Da mass shift (Ehrich et al., 2005).

4.4.6.2 Primer design

Genomic DNA sequences were downloaded from the University of California, Santa Cruz genome browser (http://www.genome.ucsc.edu). In order to maximize coverage, both the forward and reverse strand of a target region were included for amplicon design. The selected genomic sequence was subsequently exported to the primer design software MethPrimer (http://www.urogene.org/methprimer). Once genomic DNA is uploaded into this application, an *in silico* bisulfite conversion is performed to facilitate the primer design. PCR primer design criteria consist of the following: An optimal primer melting temperature of 62°C (range: 56-64°C); Primer length ranges from 20-30 nucleotides, excluding tag addition. Amplicons vary from 100-500 bp in length, with a desired length of 400 bp. All primers designed for methylation analysis using the MassCLEAVE assay are online (Gebhard et al., 2010) or will be available upon publication.

Primers were ordered in 96-well format at 100 µM concentration (Integrated DNA Technologies, California, USA or SIGMA). Prior to PCR set up, a 96-well primer mix plate (Sarstedt V-bottom, Newton, USA) was assembled, with each well containing 1 µl of both the forward and reverse primers of a primer pair and 198 µl ddH$_2$O to give a final concentration of 0.5 µM each.

4.4.6.3 Bisulfite conversion

Bisulfite treatment of genomic DNA was performed using a commercially available kit from Zymo Research Corporation (California, USA). The EZ DNA Methylation Kit^{TM} facilitates the conversion of cytosine to uracil due to the reaction that takes place between cytosine and sodium bisulfite. The conversion reaction was performed, using 1 µg of genomic DNA, according to the manufacturer's protocol, but with the following alternative conversion parameter:

Step 1:	95 °C	30 s
Step 2:	50 °C	15 min
Step 3:	Repeat steps 1-2 for 20 cycles	
Step 4:	4 °C	hold

4.4.6.4 PCR amplification

PCR master mixes were prepared in 384 well plates (ABgene) and made as follows per reaction:

Component	Volume for single reaction	Final concentration
ddH$_2$O	1.42 µl	N/A
10x HotStarBuffer	0.5 µl	1x
dNTP mix 25 mM each	0.04 µl	200 µM
5 U/µl Hot Star Taq	0.04 ml	0.2 U
DNA Template	1 µl	5-10 ng

To each reaction 2 µl primer mix was added, giving a final reaction volume of 5 µl, with the concentration of 500 pM of the forward and reverse primer. Then the plate was sealed with AB-0558 spun down, centrifuged and incubated in a Veriti 384 well thermal cycler (Applied Biosystems) with the following program:

Table 4-10 Reaction parameter for bisulfite conversion

PCR step		Cycling parameter
Initial melting		94 °C, 4 min
45 cycles	Melting	94 °C, 20 s
	Annealing	59 °C, 30 s
	Elongation	72 °C, 1 min
Final elongation		72 °C, 3 min
Cooling		4 °C, forever

4.4.6.5 Shrimp alkaline phosphatase (SAP) treatment

Unincorporated nucleotides can disturb downstream applications and are therefore enzymatically inactivated. Under alkaline conditions SAP removes phosphate groups from several substrates including deoxynucleotide triphosphates, rendering it unavailable for further polymerase catalyzed reactions. The SAP solution was prepared as follows:

Component	Volume for single reaction
RNAse free water	1.7 µl
SAP	0.3 µl

2 µl of the SAP solution was added to each PCR reaction with the 96 channel pipetting robot MassARRAY Liquid Handler and Fusio^{TM} Chip Module (Matrix). The plate was sealed with AB-0558, centrifuged and incubated as follows on a Veriti 384 well thermal cycler (Applied Biosystems):

Step 1:	37 °C	20 min
Step 2:	85 °C	5 min
Step 3:	4 °C	hold

4.4.6.6 Reverse transcription and RNA base-specific cleavage

Transcription and Cleavage were performed using a single mix containing:

Component	Volume for single reaction
RNase free water	3.21 µl
5× T7 Polymerase buffer	0.89 µl
Cleavage Mix (T mix)	0.22 µl
DTT (100 mM)	0.22 µl
T7 R&DNA Polymerase (50 U/µl)	0.4 µl

Methods

RNaseA	0.06 µl
Total volume	5 µl

5 µl of the mix and 2 µl of the SAP-treated PCR reaction were transferred into a new 386-well plate with the 96 channel pipetting robot MassARRAY Liquid Handler and Fusio^{TM} Chip Module (Matrix), sealed with AB-0558, centrifuged and incubated on a Veriti 384 well thermocycler C (Applied Biosystems) for three hours at 37 °C.

4.4.6.7 Desalting the cleavage reaction

Salt ions are co-vaporized when acquired during MALDI-TOF analysis and are therefore visible in the mass spectra. This would irritate the analysis of the mass spectra. Therefore the reactions need to be desalted. For desalting of the transcription/cleavage mix 20 µl water was added to each reaction with the MassARRAY Liquid Handler (Matrix) followed by the addition of 6 mg CLEAN resin per reaction. The plates were rotated slowly for 10 minutes and spun down to collect the resin at the bottom of the wells.

4.4.6.8 Transfer on SpectroCHIP and acquisition

The SpectroCHIP contains the matrix on which the sample probes are spotted and consists of a crystallized acidic compound. When the analyte is spotted onto the matrix its solvent dissolves the matrix, and when the solvent evaporates the matrix recrystallizes with analyte molecules enclosed in the crystals. The DNA samples are transferred on a SpectroCHIP either with the Phusio Chip Module or the 24 pin-head nanodispenser and are analyzed with the MassARRAY Compact System MALDI-TOF MS (all from Sequenom). The co-crystallized analyte is acquired with a laser while the matrix is predominantly ionized, protecting the DNA from the disruptive laser beam. Eventually, the charge is transferred to the sample and charged ions are created which are accelerated in a vacuum towards a detector that measures the particle's time of flight.

4.4.6.9 Interpretation of data output and quality control

Acquired data was processed with the EpiTYPER Analyzer software (version 1.0.5, Sequenom). The MS is calibrated with a four point calibrant (Sequenom) with 1479, 3004, 5044.4 and 8486.6 kDa particles. Relative to this calibration the accelerated analytes generate signal intensity (y-axis) versus mass (kDa, x-axis) plots. With the sequence of every amplicon known, the software can virtually process the sequence and predict the fragments from the *in vitro* transcription/RNase digestion and relocate CpG units. If expected and incoming information match, the signal intensities of the methylated and unmethylated DNA templates are compared and quantified. A normal calibrated system is able to measure fragments between a range of 1500 and 7000 Dalton. Fragments outside of this range and fragments whose mass peaks are overlapping with multiple other fragments cannot be analyzed.

As an additional control feature to assess the quality of DNA samples and the consistency of the technology employed in this study, DNA methylation values for each assay were determined in fully methylated DNA and completely unmethylated DNA. A mixed control was also assembled by combining equal quantities of fully methylated and unmethylated control DNA.

In order to generate fully unmethylated genomic DNA *in vitro*, genomic DNA was amplified using the REPLI-g Mini/Midi kit (Qiagen) according to the manufacturer's instructions. Purification of amplification products was performed using QIAamp DNA Micro Kit (Qiagen) as indicated in the manufacturer's manual. Secondary, to generate fully methylated DNA as a control for methylation analysis, genomic DNA was methylated using *Sss* I methyltransferase (see section 4.4.1).

A desired percentage of methylation was generated by mixing an appropriate amount of unmethylated (0%) and fully methylated (100%) DNA.

4.4.6.10 Calculation of EpiTYPER methylation score ratio

To compare the high resolution mass spectrometry data with intermediate resolution microarray data, we assigned each microarray probe with a so-called EpiTYPER methylation score ratio which basically represents a mean scaled $\log_{10}$ ratio of all measured CpGs in a region 300 bp upstream and downstream of a microarray probe center. EpiTYPER methylation values for individual CpGs were transformed into $\log_{10}$ ratios using the formula: $(\log_{10}((T+0.01)/M+0.01)$ where T and M represent EpiTYPER methylation ratios of the cell line and normal monocytes, respectively (0.01 was added to each value to avoid division by zero). To account for the non-linear enrichment obtained by MCIp and to adjust the EpiTYPER methylation score ratio to the range of MCIp $\log_{10}$ ratios, the $\log_{10}$ ratio of individual CpGs was corrected by an empirically determined factor that weighted for methylation strength $(\log_{10}(ABS(T-M))/2+1.01)*2)$. The EpiTYPER methylation score of a microarray probe was then calculated as the mean scaled $\log_{10}$ ratio of all measured CpGs in a region 300 bp upstream and downstream of a microarray probe center.

4.5 *De novo* motif discovery

4.5.1 Algorithm for *de novo* motif finding

Motif discovery was performed using a comparative algorithm similar to those previously described (Barash Y. et al., 2001). An in-depth description and benchmarking of the software suite HOMER (**H**ypergeometric **O**ptimization of **M**otif En**R**ichment; http://biowhat.ucsd.edu/homer/) that was developed for motif discovery will be published elsewhere (Benner et al., in preparation). Briefly, sequences were divided into target and background sets for each application of the algorithm. Background sequences are then selectively weighted to equalize the distributions of CpG content in target and background sequences to avoid comparing sequences of different sequence content. Motifs are found separately by first performing exhaustively screening all oligo sequences for enrichment in the target set compared to the background set using the cumulative hypergeometric distribution. Up to two mismatches were allowed in oligo sequences to increase the sensitivity of the method. The top 50 sequences of each length with the lowest P values were then converted into probability matrices and heuristically optimized to maximize hypergeometric enrichment of each motif in the given data set. As optimized motifs are found they are removed from the data set to facilitate the identification of additional motifs.

4.5.2 ChIP-on-chip peak calling and motif annotation

Transcription factor-bound regions were identified using a sliding window approach and the averaged data sets from two independent experiments (correlation coefficients for $\log_{10}$ ratios of replicate ChIP-on-chip experiments: $r^2_{Sp1}=0.95$; $r^2_{YY1}=0.88$; $r^2_{NRF1}=0.75$). The window included five probes with a maximal distance of 500 bp between two neighboring probes. A cumulative $\log_{10}$ ratio of ChIP/input over the five probes of 1.5 was used as a lower threshold for detecting a binding event. To study the correlation between motif presence and actual transcription factor binding, we annotated each motif for Sp1, NRF1 and YY1 with mean signal intensity ratios ($\log_{10}$) of all microarray probes from the corresponding ChIP-on-chip experiments in the range of ± 150 bp around it. The lower limit for binding of a motif in normal blood monocytes was set at a mean signal intensity $\log_{10}$ ratio of 0.4. A motif was grouped as bound if both, the sliding window approach and the motif centered approach indicated binding. A motif was grouped as not bound if both methods indicated non-binding.

5 Results

Due to space limitations, this section only contains parts or summaries of the expression, MCIp-on-chip and MassARRAY data. Complete figures, tables, lists and UCSC tracks are or will be available online within the supplementary information of the corresponding publications.

5.1 Detection of methylated DNA by methyl-CpG immunoprecipitation (MCIp)

To date, the investigation of aberrant CpG island methylation has primarily taken a candidate gene approach. However, in order to assess the clinical potential of hypermethylation profiles and to identify relevant marker genes, methods for the genome-wide detection of hypermethylation are required. Because there were no suitable methods available, we developed a sensitive approach in our lab that enabled the detection of methylated CpG dinucleotides using only very little DNA quantities but which still allows for global detection of DNA methylation.

The basis for detection of methylated DNA is provided by a recombinant antibody-like fusion protein that consists of the human methyl-CpG binding domain 2 (MBD2), a flexible linker polypeptide and the Fc-proportion of human IgG1. Design and generation of the MBD-Fc protein is described in section 4.2 (for further details see also Gebhard, 2005 and Gebhard et al., 2006). In previous studies performed in our own lab it could be shown that the MBD-Fc protein is able to bind methylated DNA in an antibody-like manner. As previously shown, *in vitro* methylated or unmethylated PCR fragments with different CpG density could be detected by MBD-Fc on nylon membranes in a linear fashion and dependent on the 5mC content (Gebhard, 2005 and Gebhard et al., 2006b).

Based on the recombinant MBD-Fc a novel technique was developed in our lab that enables the unbiased genome-wide detection of CpG methylation, the so-called methyl-CpG immunoprecipitation (MCIp). The approach allows for the detection of the methylation status of specific CpG island promoters (in combination with real-time PCR) and also allows for the generation of genome-wide promoter methylation profiles (in combination with microarray or next generation sequencing technology).

In contrast to another recently developed immunoprecipitation approach using 5-methyl specific antibodies (called MeDIP or mDIP) that specifically enriches for methylated fragments (Weber et al., 2005), MCIp can divide the bulk of genomic DNA fragments into separate fractions of increasing methylation density. This is due to the fact, that MBD-Fc recognizes the hydration of methylated DNA rather than 5mC itself (Ho et al., 2008). Therefore using increasing salt concentrations allows for the fractionation of genomic DNA fragments according to their methylation degree (Gebhard et al., 2006b; Gebhard et al., 2006a; Schilling and Rehli, 2007). A schematic representation of the MCIp approach is shown in Figure 5-1.

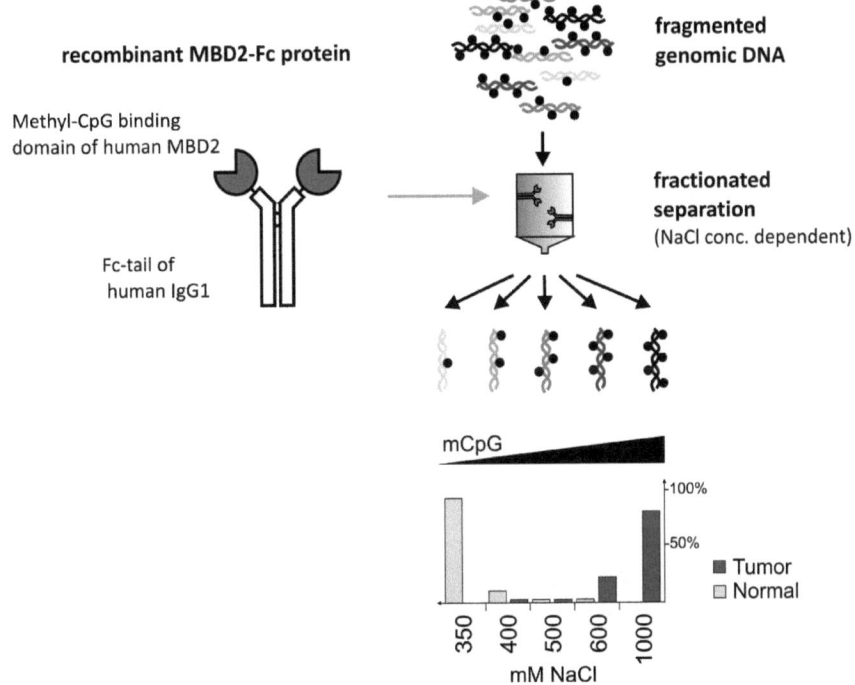

Figure 5-1 Schematic presentation of the methyl-CpG immunoprecipitation approach (MCIp)
Fragmented genomic DNA is incubated with saturating amounts of MBD-Fc Protein A-Sepharose matrix at a low NaCl concentration. The column is spin-washed with increasing salt concentrations leading to fractionation of the fragments according to the methylation density. The flow-through consists of fragments with little or no CpG methylation, while high salt fractions contain strongly methylated and CpG-rich fragments.

5.1.1 Detection of *in vitro* methylated DNA promoter fragments

An initial characterization of the MBD-Fc protein and its ability to bind CpG-methylated DNA had already been done before starting this dissertation (Gebhard, 2005; Gebhard et al., 2006b). To further describe the properties of MBD-Fc, we also tested whether this approach allows for the detection of different degrees of methylation for a single gene locus. Therefore a CpG island fragment (CPM) was cloned into the CpG-free vector pCpG-mcs, linearized and *in vitro* methylated using increasing amounts of the methyl donor S-adenosylmethionine to obtain fragments with varying methylation densities (see section 4.4.2). Subsequently, the fragments with different methylation levels were combined, fractionated by MCIp and subjected to bisulfite sequencing (see section 4.4.3). If genomic DNA is treated with sodium bisulfite, unmethylated cytosines are deaminated into uracil and transformed into thymidine residues during PCR, whereas methylated cytosines still appear as cytosines after amplification (Frommer et al., 1992b). Figure 5-2 demonstrates that partially *in vitro* methylated DNA fragments can be separated according to their methylation degree in the developed methyl-CpG immunoprecipitation approach using increasing salt concentrations.

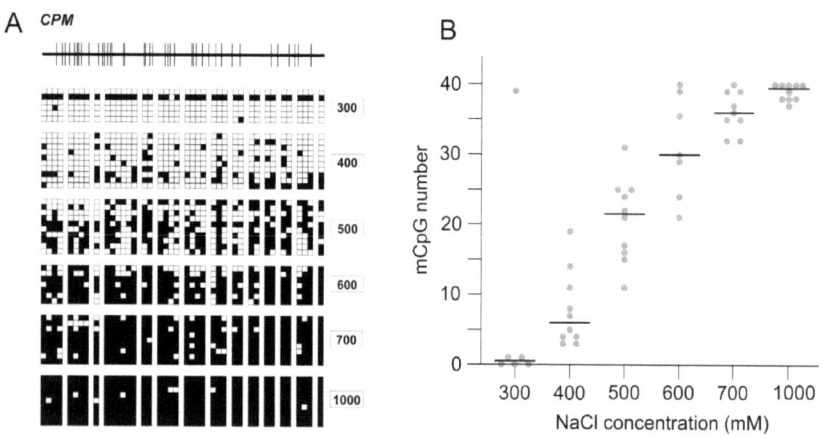

Figure 5-2 Bisulfite sequences of an *in vitro* partially methylated gene locus after MCIp
A mixture of *Sss* I-methylated CPM CpG island promoter fragments (schematic representation on the top of A) with varying methylation density was fractionated using MCIp. Fragments were recovered from each fraction, subjected to bisulfite treatment and cloned. Several independent inserts were sequenced. Results are represented schematically. (A) Squares mark the position of CpG dinucleotides (empty, unmethylated; filled, methylated). 300-1000 indicates the salt concentration (mM) used to elute the different fragments. (B) The results are represented as a graph where each point represents the number of methylated CpGs in one clone. Horizontal bars represent the median number of methylated CpG dinucleotides at one specific salt concentration.

This and previous test experiments performed with specific methylated and unmethylated DNA fragments showed that the recombinant MBD-Fc fusion protein binds CpG-methylated DNA. The binding capacity is contingent on the NaCl concentration as well as on the CpG methylation density of the bound DNA.

5.1.2 Detection of methylated genomic DNA fragments

To test, whether the MCIp procedure could be applied to discriminate methylated and unmethylated DNA fragments from genomic DNA, the newly developed approach was used to precipitate *Mse* I-restricted genomic DNA. *Mse* I was chosen for DNA fragmentation, because it is known to preferentially cut in regions of low CpG content while leaving many CpG islands uncut (Cross et al., 1994).

5.1.2.1 Combination of MCIp and real-time PCR to detect the methylation status of specific CpG island promoters

To explore this type of application on a single gene level, *Mse* I-restricted genomic DNA of *in vitro Sss* I-methylated and untreated normal DNA from monocytes of a healthy donor were subjected to MCIp. The enrichment of four different CpG island promoters and one promoter with low CpG density in the different fractions was determined relative to the input DNA using LightCycler real-time PCR. As a positive control the *SNRPN* CpG island promoter was used. This gene is subject to maternal imprinting with one of its two copies being methylated also in normal cells (Zeschnigk et al., 1997). As expected, in normal DNA, the two differentially methylated allele-fragments were enriched in two separate fractions. The unmethylated allele-fragment was mainly eluted at 400 mM NaCl, whereas the methylated allele fragment was eluted at 1000 mM NaCl. However, with *Sss* I methylated DNA only one positive elution fraction could be observed because both alleles are methylated and were detected in the 1000 mM NaCl fraction (Figure 5-3A). In the case of the *CDKN2B* gene (also known als $p15^{INK4b}$) the promoter fragment was mainly recovered in the low salt fractions from normal DNA and in the high salt fraction from *Sss* I-methylated DNA (Figure 5-3B). Similar results were obtained for the human estrogen receptor 1 (*ESR1*) gene and the human Toll-like receptor 2 gene (*TLR2*) (data not shown). Another test locus used was the promoter fragment of the *CHI3L1* gene. This gene fragment however, showed different elution profiles: the untreated normal fragment was mostly detected at low salt concentrations (400 and 500 mM). When the DNA was *in vitro* methylated only a slight shift towards higher NaCl concentrations was observed (Figure 5-3C). This is due to the lower CpG density of the *CHI3L* promoter. The detected fragment only contains 12 CpG dinucleotides and the difference between methylated and unmethylated fragment is only five to six methylated CpG residues. Together, these data show that the MCIp approach is able to discriminate even slight differences in CpG methylation.

Analysis of elution profiles shown in Figure 5-3 suggests that (i) a 200- to 300-fold enrichment of stronger over less methylated genomic fragments can be obtained in either low or high salt fractions, (ii) fragments with low CpG density are largely excluded from the high salt fraction, and (iii) the fractionated MCIp approach may allow for the resolution of relatively small differences in CpG methylation density.

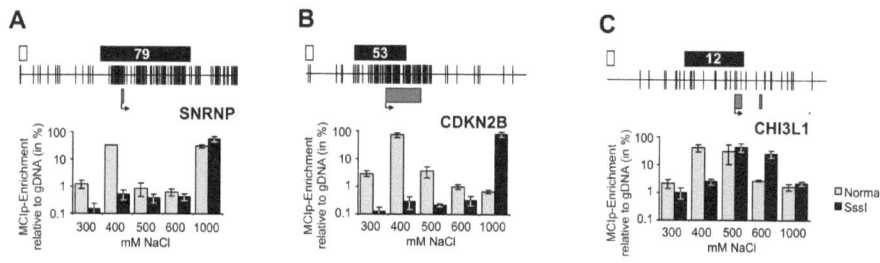

Figure 5-3 MCIp detection of CpG methylation in specific CpG island promoters using real-time PCR
Fractionated MCIp was used in combination with LightCycler real-time PCR to detect the methylation status of specific CpG island promoter fragments from untreated (grey bars) and *Sss* I-methylated (black bars) *Mse* I-restricted genomic DNA fragments. Recovered gene fragments from MCIp eluates (different salt concentrations are indicated) and an equivalent amount of input DNA are amplified using LightCycler real-time PCR. Values (mean ± SD, n=4 using at least two different preparations of MBD-Fc) of the different fractions represent the percentage of recovery and are calculated relative to the amount of the respective input DNA (100%). Above each figure a 3 kb region of the corresponding CpG island is schematically represented. Each CpG dinucleotide is represented by a vertical line. Black boxes on the top indicate the position of the *Mse* I fragments that are detected with the number indicating the number of CpG dinucleotides within the fragment. The positions of exons are indicated as dark grey boxes and transcription start sites by an arrow. The white box represents a 100 bp fragment.

In the next step it was determined whether the fractionating MCIp approach is able to detect aberrant hypermethylation in tumor samples. *Mse* I-digested genomic DNA from three leukemia cell lines KG-1 (acute myeloid leukemia), U937 (histiocytic lymphoma, monocytic), THP-1 (acute monocytic leukemia) as well as from monocytes of a healthy donor were subjected to MCIp. Using LightCycler real-time PCR the enrichment of four different CpG island promoters in the 1000 mM NaCl fraction was detected relative to the input DNA. The imprinted *SNRPN* gene promoter was used as a positive control. Another test locus, the promoter of the human Toll-like receptor 2 gene (*TLR2*) was chosen, because our group had previously shown that this promoter fragment is strongly methylated in U937 cells, but not in THP-1 cells (Haehnel et al., 2002). The promoters of the human estrogen receptor 1 (*ESR1*) (Dodge et al., 2001) gene and the human *CDKN2B* ($p15^{INK4b}$) (Chim et al., 2003; Dodge et al., 1998; Haehnel et al., 2002) gene were used because they are known to be frequently methylated in leukemic cells. The results are summarized in Figure 5-4. The *SNRPN* gene promoter was significantly enriched in all leukemia cell lines as well as in normal cells which is in concordance with its imprinting-related methylation status. The *TLR2* locus was enriched and therefore methylated in KG-1 and U937 cells, but not in THP-1 or normal cells. The methylation pattern of the *TLR2* promoter fragment was confirmed by bisulfite sequencing (Haehnel et al., 2002). The results for *ESR1* (KG-1) and *CDKN2B* (KG-1 and U937) were also in line with previously published studies (Chim et al., 2003; Dodge et al., 2001; Gebhard, 2005; Issa et al., 1996). In THP-1 cells the PCR amplification of the *CDKN2B* promoter fragment failed due to a deletion of this locus. None of the three *Mse* I fragments (with an exception of the imprinted *SNRPN* gene locus) were significantly enriched in the DNA from normal monocytes which correlates with the normally unmethylated state. From these results it can be concluded that MCIp fractionates genomic DNA according to the degree of methylation and specifically enriches strongly methylated DNA fragments in the high salt fraction.

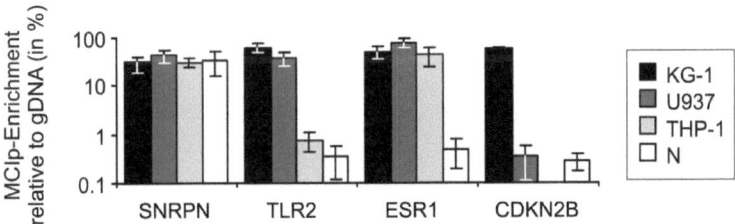

Figure 5-4 MCIp detection of CpG island methylation in specific CpG island promoters using real-time PCR
SNRPN, *TLR2*, *ESR1*, and *CDKN2B* gene fragments in the high salt fraction of three human myeloid leukemia cell lines (KG-1, U937, THP-1) as well as normal blood monocytes (N) were analyzed using real-time PCR as described in Figure 5-3.

Results

5.1.2.2 Sensitivity and linearity of the MCIp approach

To test the sensitivity of the fractionation approach, decreasing amounts of *Mse* I-treated U937 DNA were subjected to MCIp. The enrichment of the *TLR2* (strong methylation in U937) and *CDKN2B* gene fragments (no methylation in U937) were determined by LightCycler real-time PCR. Figure 5-5A shows that a significant enrichment of the *TLR2* fragment could be achieved using as little as 1 ng DNA, which corresponds approximately 150 tumor cells.

Samples derived from tumors may contain a specific and variable number of normal cells that would be expected to be unmethylated at most CpG islands. To test the linearity of the MCIp approach with respect to cell purity, mixtures of *Mse* I-treated DNA from normal monocytes and the leukemia cell line KG-1 were fractionated using increasing salt concentrations and again analyzed by real-time PCR with specific primers for the *TLR2* locus (methylated in KG-1 and unmethylated in normal blood cells). As shown in Figure 5-5B the *TLR2* fragment was only detected in samples containing KG-1 DNA and the signal increased gradually with increasing amount of KG-1 cells in the mixture.

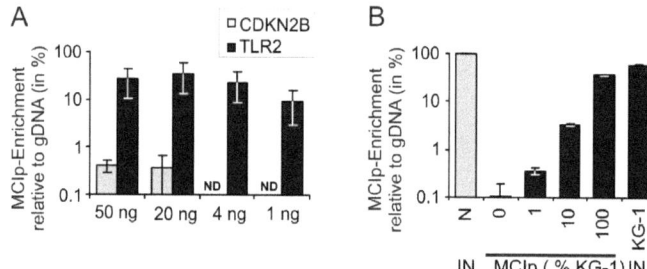

Figure 5-5 Sensitivity (A) and linearity (B) of the MCIp approach
(A) Decreasing amounts of *Mse* I-treated U937 DNA were subjected to MCIp. *TLR2* and *CDKN2B* gene fragments were analyzed by LightCycler real-time PCR as described in Figure 5-3. (B) MCIp was performed with mixtures of *Mse* I-treated normal blood monocytes (N) and increasing amounts of KG-1 cells. *TLR2* enrichment in the 1000 mM salt fraction was determined by LightCycler real-time PCR.

As demonstrated above, MCIp combined to real-time PCR was very sensitive. However, for early diagnosis as well as methylation detection from body fluids and trace amount analysis in the post therapy, a method for an ultrasensitive methylation detection may be necessary and suggestive. Because MassARRAY Quantitative Gene Expression (QGE) provides orders of magnitude greater sensitivity than real-time quantitative PCR, and permits very closely related genes to be assayed reliably and quantitatively, MCIp was combined with QGE. This type of assay could provide an alternative method for the methylation specific PCR (MSP) which suffers from many disadvantages. Firstly, MSP is based on bisulfite treatment, secondly it requires very extensive validation, thirdly it is only an indirect measurement of DNA methylation and finally, it has only limited possibilities for quantitation. Using MCIp and subsequent MassARRAY (QGE), we wanted to circumvent the disadvantages of the MSP method and create a system to detect methylated DNA in an ultrasensitive and reproducible manner.

In the QGE assay, quantitation is based on a competitive PCR in which a cDNA template of interest and a competitor (internal standard) are co-amplified in the same reaction. Each competitor molecule matches its target cDNA sequence at all nucleotide positions except a single base so the two can be resolved using mass spectrometry-based genotyping. Titration of competitor concentrations is used to determine the competitor concentration at which amplification between cDNA and competitor is equal. This point, termed the EC50, is determined by plotting cDNA allele frequency vs. competitor concentration. Non-linear regression is used to calculate the point at which the cDNA and competitor are at a 1:1 ratio. Each assay uses a single base extension to distinguish between the target cDNA and competitor template (for more detailed information see section 4.3.4.4, primers and competitors are listed in section 3.6.7).

First, it was tested whether MCIp combined with QGE is able to detect the correct number of copies of methylated and unmethylated DNA fragments subjected to MCIp in the lower or higher salt fractions, respectively. As a test system, the *MGMT* (O-6-methylguanine-DNA methyltransferase) locus of unmethylated DNA from normal monocytes as well as the same locus of *in vitro* methylated monocyte DNA was used. In a first step, both unmethylated and methylated DNA from monocytes were restricted using *Msp* I. Afterwards 1,500 copies of unmethylated DNA and methylated DNA, respectively, and in addition two times a 50:50 mixture (750 copies of each, methylated and unmethylated DNA) were subjected to MCIp. DNA fragments were separated using increasing salt concentrations (200, 300, 350, 400, 450, 500, 600 and 1000 mM NaCl) and the different fractions were analyzed for enrichment of the *MGMT* gene locus using QGE and primers as well as competitors specific for the *MGMT Msp* I-restricted gene fragment. As illustrated in Figure 5-6 MCIp is able to detect the copies of unmethylated as well as methylated DNA with high specificity and accuracy.

Results

The MassARRAY system allows for the detection of almost all DNA fragments from the input DNA. 70-80% of the unmethylated DNA fragments were detected in the lower salt concentrations. A similar percentage of methylated DNA fragments were recovered and detected in the high salt fractions. The remainder of the DNA probably got lost during the procedure.

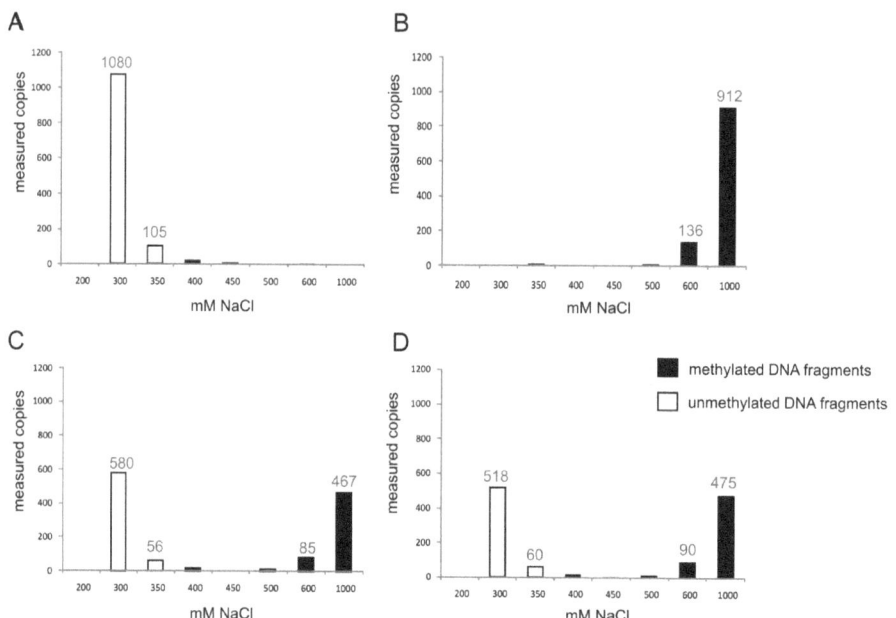

Figure 5-6 MCIp detection of the *MGMT* locus using quantitative gene expression (QGE)
Fractionated MCIp was used in combination with QGE to detect the methylation status of the *MGMT* locus from untreated (grey bars) and *Sss* I-methylated (black bars) *Msp* I-restricted genomic DNA fragments. In (A) 1500 copies of the unmethylated, in (B) 1500 copies of the methylated genome and in (C) and (D) 750 copies of the unmethylated genome mixed with 750 copies of the methylated genome were subjected to MCIp. Recovered gene fragments from MCIp eluates (different salt concentrations are indicated) were quantified using QGE. Values of the different fractions represent the measured number of copies and were calculated related to the competitor concentration using the EC50 value. Above each figure the exact number of detected copies is represented.

To test the sensitivity of the approach, a 10% mixture of DNA (150 copies of methylated DNA together with 1363 copies of unmethylated DNA) and a 1:2, 1:4 and 1:8 dilution of the same mixture were subjected to MCIp and subsequently to QGE. 16 replicates proofed the high reproducibility of the approach. After inactivating the outliers, the values were averaged. As shown in Figure 5-7 the method allows the detection of as little as 24 copies of methylated DNA. Furthermore the graph illustrates again the accuracy of the method: the correct copy number of unmethylated and methylated DNA fragments is detected dependent on the degree of dilution in a linear fashion ($r^2=0.99$) in a range of more than three logarithmical stages.

A future aim would be to further improve this method and to achieve a multiplexing for high-throughput screening of patient samples for risk assessment.

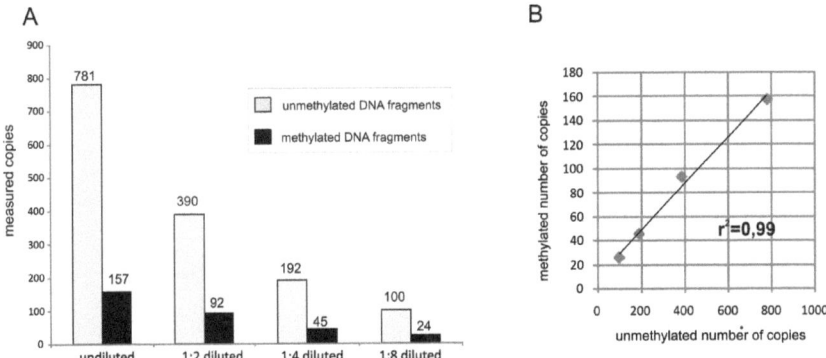

Figure 5-7 Sensitivity and linearity of the MCIp approach combined to QGE
(A) MCIp was performed with decreasing amounts of *Msp* I-treated 10% mixtures (1363 copies unmethylated monocyte DNA mixed with 150 copies *Sss* I-methylated DNA). Enrichment of the *MGMT* gene fragments (the unmethylated fragments in the lower salt concentrations, the methylated ones in the higher salt concentrations) were analyzed by QGE. (B) Correlation between methylated and unmethylated copy numbers within the dilution series of a 10% mixture ($r^2=0.99$). Values are mean ± SD (n=4).

5.2 Combination of MCIp and 12K CpG island microarray analysis

Data from this section have been published in the journal *Cancer Research*. Microarray data were deposited with GEO (gene expression analyses: GSE 3280; comparative MCIp hybridizations: GSE).

To achieve a genome-wide identification of aberrant methylation patterns, the MCIp approach was combined with microarray technology. For unbiased genome-wide analysis of aberrant methylation profiles, the MCIp technique had to be adapted to the microarray technology and a series of optimization steps in terms of amplification, labeling as well as hybridization were already performed previously (Gebhard, 2005).

Results

Methyl-CpG immunoprecipitations were performed with 300 ng *Mse* I-restricted DNA from three leukemia cell lines (KG-1, U937, THP-1) as well from normal blood monocytes. The high salt fractions (600-1000 mM) containing the strongly methylated CpG island promoter fragments were spin purified and afterwards amplified using ligation-mediated PCR. The resulting amplicons were directly labeled with Cy5-dCTP (normal DNA) and Cy3-dCTP (tumor DNA) using the Exo-Klenow enzyme. Subsequently each leukemia sample was cohybridized with the normal control sample to CpG island microarrays (UHN Microarray Centre, Toronto, Canada). Figure 5-8 represents the schematic workflow of the procedure for DNA methylation profiling.

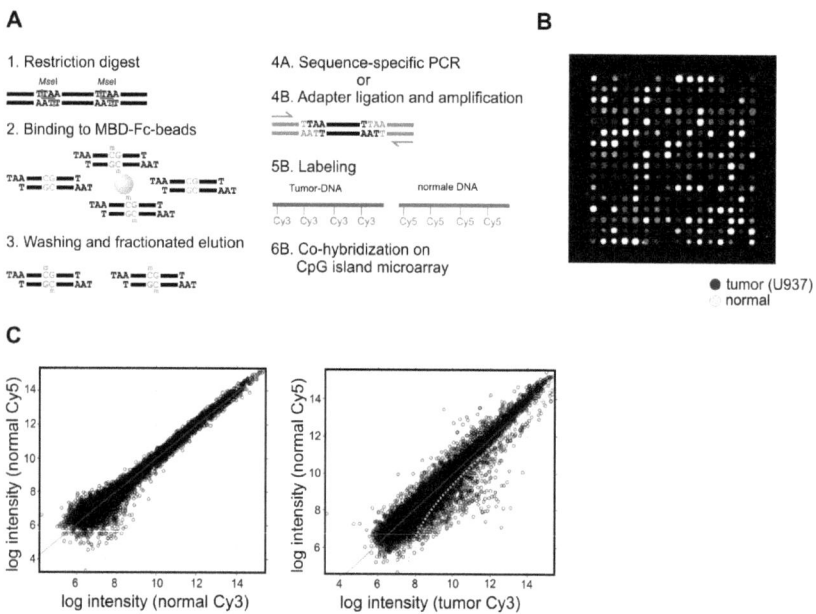

Figure 5-8 Schematic representation of DNA methylation profiling using MCIp and CpG island microarrays
(A) Global methylation levels were determined using MCIp (steps 1-3) and subsequent cohybridization of amplified tumor (Cy3-labeled) and normal DNA (Cy5-labeled) on microarrays (steps 4-6). (B) Frequently methylated U937 DNA shows a higher fluorescence for Cy3 (dark color) and normal DNA for Cy5 (grey color). (C) Representative two-dimensional scatter plots are shown for a control hybridization experiment of human blood monocytes DNA (normal/normal) (left) and a hybridization displaying the differentially methylated CpG fragments between U937 (tumor) and human blood monocytes (normal) (right).

The CpG island array contains 12,192 CpG island clones from a *Mse* I-CpG DNA library that was originally prepared by MeCP2-column purification of non-methylated CpG island fragments (Cross et al., 1994). Representative scatter plots of microarray hybridizations are presented in Figure 5-8C. On the left side the scatter plot for the control hybridization experiment is shown: Cy3-labeled normal monocytes are plotted against Cy5-labeled normal monocytes. On the right side the Cy3-labeled U937 tumor sample is plotted against Cy5-labeled normal monocytes. Comparing the two plots, differently methylated CpG fragments between the U937 tumor cell line and normal human blood monocytes are displayed. Signals corresponding to both hypo- and hypermethylated fragments in the tumor sample were observed. This work focused on the analysis of hypermethylated fragments. To identify possible tumor suppressor genes or other marker genes that are affected by hypermethylation, results of three independent MCIp experiments (using two different MBD-Fc preparations and three independent DNA preparations) were analyzed in conjunction. Hybridization signals that were more than twofold enriched in the leukemia sample and consistently different in at least one cell line were selected for further analysis. In total, THP-1, U937 and KG-1 cells showed 277, 454 and 330 differential hybridization signals, respectively. 191 out of 535 spots analyzed were unambiguously annotated and located within close proximity (approximately ± 3,000 bp) to predicted transcriptional start sites and were chosen for further analysis. Since some sequences were represented more than once on the CpG island microarray, the final, non-redundant list of differentially methylated DNA fragments contained 131 entries that were in close proximity of 134 genes (Table 5-1). The hypermethylated genes shown in Table 5-1 are involved in many important biological functions. Most strikingly, half of the genes with an assigned biological function (46 of 89) are involved in DNA binding and transcriptional regulation. Nine of the listed genes have been previously identified as hypermethylation targets in cancer: *LMX1A* (Paz et al., 2003), *TFAP2A* (Douglas et al., 2004), *CR2* (Schwab and Illges, 2001), *DCC* (Sato et al., 2001), *MYOD1* (Jones et al., 1990), *DLEC1* (Yuan et al., 2003), *AKAP12* (Choi et al., 2004), *SSIAH2 (LOC28314)* and *FOXF1* (Weber et al., 2005).

Table 5-1 Hypermethylated gene fragments in myeloid leukemia cell lines

Gene			CpG-methylation				mRNA expression				
Name	Symbol	Location	KG1	U937	THP1	Positon	KG1	U937	THP1	N	Probe Set ID
hypothetical gene	LOC400027	12q12	1.82	1.16	1.19	down	NC/P	NC/P	NC/P	A	226413_at
branched chain aminotransferase 2, mitochondrial	BCAT2	19q13	1.24	1.18	1.29	proximal	1	1.8	NC/P	P	203576_at
chondrolectin	CHODL	21q11.2	1.52	1.49	1.44	down	NC/A	NC/A	NC/A	A	219867_at
collagen, type XIV, alpha 1 (undulin)	COL14A1	8q23	2.02	2.97	2.03	down	NC/A	NC/A	NC/A	A	1562189_at
cytochrome P450, family 27, subfamily B, polypeptide 1	CYP27B1	12q13.1-q13.3	1.55	1.87	2.15	down	NC/A	NC/P	0.8	A	205676_at
v-erb-a erythroblastic leukemia viral oncogene homolog 4	ERBB4	2q33.3-q34	1.63	1.34	1.36	down	NC/A	NC/A	NC/A	A	241581_at
family with sequence similarity 5, member B	FAM5B	1q24.1-q25.3	2.55	1.32	1.92	proximal	NC/A	NC/A	NC/A	A	214822_at
fibroblast growth factor 12	FGF12	3q28	1.20	1.76	1.52	proximal/down	NC/A	NC/A	NC/A	A	240067_at
hypothetical gene	FLJ13192	15q14	1.79	1.49	1.37	down	NC/A	NC/A	NC/A	A	233382_at
hypothetical gene	FLJ20366					up/down	NC/A	NC/A	NC/A	A	218692_at
hypothetical gene	FLJ36633	8q23.2	1.90	1.39	1.45	up	ND	ND	ND	ND	NA

Results

Gene			CpG-methylation				mRNA expression				
Name	Symbol	Location	KG1	U937	THP1	Positon	KG1	U937	THP1	N	Probe Set ID
hypothetical protein	FLJ20972	1p34.2	1.46	1.29	1.82	down	NC/M	NC/P	NC/P	P	230897_at
hypothetical protein	FLJ35074					down	NC/A	1.6	NC/A	A	1560503_a_at
transcription factor AP-2 alpha	TFAP2A	6p24	2.07	1.09	1.59	up	NC/A	2.4	NC/A	A	204653_at
laeverin	FLJ90650	5q23.1	2.14	3.00	2.77	up/down	NC/A	NC/A	NC/A	A	235382_at
forkhead box F1	FOXF1	16q24	1.40	1.84	2.72	down	2.7	NC/A	NC/A	A	205935_at
glycoprotein M6A	GPM6A	4q34	1.30	2.05	1.19	down	NC/A	NC/A	NC/A	A	209469_at
GS homeobox 2	GSH2	4q11-q12	1.34	2.29	1.26	down	NC/A	NC/A	NC/A	A	230338_x_at
hypocretin (orexin) receptor 2	HCRTR2	6p11-q11	1.06	1.46	1.62	down	NC/A	NC/A	NC/A	A	207393_at
Hey-like transcriptional repressor	HELT	4q35.1	1.37	3.51	1.52	up	ND	ND	ND	ND	NA
homeo box C10	HOXC10	12q13.3	1.13	1.62	1.01	up	NC/A	0.8	NC/P	A	214562_at
iroquois homeobox protein 1	IRX1	5p15.3	2.20	2.32	2.07	up	NC/A	NC/A	NC/A	A	230472_at
hypothetical protein	KIAA1024	15q25.1	1.04	2.06	1.73	down	NC/A	NC/A	NC/A	A	215081_at
LIM homeobox transcription factor 1, alpha	LMX1A	1q22-q23	2.42	2.08	2.29	down	NC/A	NC/A	NC/A	A	1553541_at
similar to seven in absentia 2 (SSIAH2)	LOC283514	13q14.13	2.43	2.73	1.54	proximal	NC/A	NC/A	NC/A	A	1560676_at
hypothetical protein	MGC12982					up	NC/P	NC/A	NC/A	A	207653_at
forkhead box D2	FOXD2	1p33	2.17	2.36	1.92	down	NC/P	NC/A	NC/A	A	224457_at
hypothetical protein	MGC42090	7p21.1	2.62	4.03	1.97	proximal	NC/A	NC/A	NC/A	A	1552293_at
hypothetical protein	MGC4767	12q24.31	1.68	2.58	2.23	proximal	1	1.2	NC/P	P	223114_at
myogenic differentiation 1	MYOD1	11p15.4	2.49	2.57	1.14	down	NC/A	NC/A	NC/A	A	206657_s_at
NK2 transcription factor related, locus 3 (Drosophila)	NKX2-3	10q24.2	2.96	4.06	1.30	down	NC/A	NC/A	3.3	A	1553808_a_at
one cut domain, family member 1	ONECUT1	15q21.1-q21.2	1.58	2.64	2.49	up	NC/A	NC/A	NC/A	P	210745_at
protocadherin gamma subfamily B, 1	PCDHGB1	5q31	1.12	1.69	3.19	down	ND	ND	ND	ND	NA
phospholipase A2, group VII	PLA2G7	6p21.2-p12	1.39	1.36	2.27	proximal	-8.9	-5.1	-5.7	P	206214_at
phospholipase D family, member 5	PLD5	1q43	1.43	1.69	1.92	down	NC/A	NC/P	NC/A	A	1563933_a_at
scinderin	SCIN	7p21.3	1.52	1.42	1.15	proximal/up	NC/A	NC/A	NC/A	A	239365_at
SLIT and NTRK-like family, member 3	SLITRK3	3q26.1	2.56	1.42	2.56	down	NC/A	NC/A	NC/A	A	206732_at
Sp5 transcription factor	SP5	2q31.1	1.30	1.89	1.00	down	NC/A	NC/A	NC/A	A	235845_at
transcription factor AP-2 gamma	TFAP2C	20q13.2	1.60	1.02	1.07	up	NC/P	NC/A	NC/A	A	205286_at
transmembrane protein 39A	TMEM39A	3q13.33	1.61	1.48	1.64	prom	NC/P	NC/P	0.6	P	222690_s_at
zinc finger protein 483	ZNF483	9q31.3	1.98	1.71	2.43	down	NC/P	NC/A	NC/A	A	1570534_a_at
zinc finger protein 565	ZNF565	19q13.12	2.08	1.74	1.80	down	NC/P	NC/A	0.9	P	228305_at
hypothetical gene	AF086288	9p24	1.75	1.34	-0.52	proximal	NC/A	NC/A	NC/A	A	237421_at
hypothetical gene	AY358245	15q24	1.02	1.33	0.01	proximal	ND	ND	ND	ND	NA
hypothetical protein	BC026095	11q12.1	1.40	1.26	-0.11	down	NC/A	NC/A	NC/A	A	1570068_at
bone morphogenetic protein 4	BMP4	14q22-q23	2.07	1.10	0.82	down	NC/A	NC/A	NC/A	A	211518_s_at
chromosome 16 open reading frame 45	C16orf45	16p13.11	1.00	1.17	-0.23	proximal	NC/A	NC/A	NC/A	A	239971_at
chromosome 1 open reading frame 126	C1orf126	1p36.21	1.84	1.44	-0.40	proximal	ND	ND	ND	ND	NA
chromosome 20 open reading frame 39	C20orf39	20p11.21	1.08	1.70	0.08	down	NC/P	NC/A	NC/A	A	231619_at
calponin 1, basic, smooth muscle	CNN1	19p13.2-p13.1	1.31	1.27	-0.37	down	NC/A	NC/A	NC/A	A	203951_at
hypothetical gene	CR611340	6p22.1	2.15	1.70	0.38	proximal	ND	ND	ND	ND	NA
chemokine (C-X-C motif) ligand 5	CXCL5	4q12-q13	1.22	1.35	0.84	proximal	NC/A	NC/A	NC/A	A	207852_at
cytochrome P450, family 1, subfamily B, polypeptide 1	CYP1B1	2p21	1.25	2.83	0.90	down	1.7	-6.6	-2.1	P	202435_s_at
fatty acid desaturase 3	FADS3	11q12-q13.1	1.01	1.15	-0.13	up	NC/A	NC/A	NC/A	A	204257_at
hypothetical protein	FLJ42262	8q12.3	2.67	2.22	0.96	up	NC/A	NC/A	-3.4	A	242193_at
homeo box D10	HOXD10	2q31.1	1.91	1.70	0.72	up	-1.5	-1.2	NC/A	A	229490_at
hypothetical protein	KIAA1465	15q24.1	1.34	1.49	0.30	up/down	NC/A	NC/A	NC/A	A	232208_at
Kruppel-like factor 5	KLF5	13q22.1	1.82	2.09	0.29	up	-3.7	NC/A	NC/A	A	209212_s_at
ladybird homeobox homolog 1 (Drosophila)	LBX1	10q24	1.03	1.64	0.97	up	NC/A	NC/A	NC/A	A	208380_at
LIM homeobox 9	LHX9	1q31-q32	2.65	1.60	0.59	up/down	NC/A	NC/A	NC/A	A	1565407_at
hypothetical protein	LOC282992	10q24.32	1.52	2.34	0.60	down	NC/A	NC/A	NC/A	A	244209_at
myeloid leukemia factor 1	MLF1	3q25.1	1.68	1.79	0.23	proximal	-2.9	-1.3	2	P	204784_s_at
5'-nucleotidase, cytosolic IA	NT5C1A	1p34.3-p33	1.53	1.57	0.80	up	NC/A	NC/A	NC/A	A	242525_s_at
phosphodiesterase 4B, cAMP-specific	PDE4B	1p31	1.64	1.01	-0.28	proximal	-4.7	-3.6	-4.8	P	211302_s_at
properdin P factor, complement	PFC	Xp11.3-p11.23	1.31	1.79	0.44	down	-5.9	-7.8	-8.2	P	206380_s_at
retina and anterior neural fold homeobox	RAX	18p11.32	1.54	1.27	-0.26	down	NC/A	NC/A	NC/A	A	208242_at
RGM domain family, member A	RGMA	15q26.1	1.33	1.92	0.46	up	-0.9	-0.8	NC/A	A	223468_s_at
Rap2-binding protein 9	RPIB9	7q21.12	1.31	1.68	0.20	up	6.7	NC/A	NC/A	P	215321_at
SHC (Src homology 2 domain containing) family, member 4	SHC4	15q21.1-q21.2	1.45	1.48	-0.11	down	NC/A	NC/A	NC/A	A	230538_at
SET binding protein 1	SETBP1	18q21.1	1.47	1.75	-0.07	up	-0.9	-0.7	-2.8	P	205933_at
SRY (sex determining region Y)-box 9	SOX9	17q24.3-q25.1	1.49	1.00	0.71	up	NC/A	NC/A	NC/A	A	202935_s_at
transcription factor 2, hepatic	TCF2	17cen-q21.3	1.99	2.92	-0.35	up	NC/A	NC/A	NC/A	A	205313_at
ELAV (embryonic lethal, abnormal vision, Drosophila)-like 2	ELAVL2	9p21	1.15	0.61	1.55	up/down	NC/A	NC/A	NC/A	A	1560905_at
forkhead box A1	FOXA1	14q12-q13	1.03	0.92	1.47	up	NC/A	NC/A	NC/A	A	204667_at
potassium channel, subfamily T, member 2	KCNT2	1q31.3	2.52	0.63	1.45	up	NC/A	NC/A	NC/A	A	244455_at
multiple PDZ domain protein	MPDZ	9p24-p22	2.14	0.93	1.79	down	NC/A	NC/A	NC/A	A	213306_at
paired box gene 9	PAX9	14q12-q13	1.41	0.59	1.48	up	NC/M	NC/A	NC/A	A	207059_at
serum deprivation response	SDPR	2q32-q33	1.19	0.52	1.61	down	1.9	-5.4	-1.9	P	222717_at
complement component (3d/Epstein Barr virus) receptor 2	CR2	1q32	1.45	0.55	-0.17	down	NC/A	NC/A	NC/A	A	244097_at
hypothetical protein	FLJ40542	22q11.21	1.36	0.43	0.60	down	-1.5	-1	-2.8	P	1556072_at

Results

Gene			CpG-methylation				mRNA expression				
Name	Symbol	Location	KG1	U937	THP1	Positon	KG1	U937	THP1	N	Probe Set ID
glial cell derived neurotrophic factor	GDNF	5p13.1-p12	1.35	0.37	0.04	up/down	NC/A	NC/A	NC/A	A	221359_at
Kruppel-like factor 11	***KLF11***	2p25	1.79	0.86	-0.18	up	-3.8	-3.1	-1.4	P	218486_at
LIM homeobox 4	LHX4	1q25.2	2.26	0.89	0.51	down	NC/A	NC/P	NC/A	A	1553157_at
zinc finger protein 215	ZNF215	11p15.4	1.82	0.78	0.43	proximal	NC/A	NC/A	NC/A	A	1555510_at
A kinase (PRKA) anchor protein (gravin) 12	AKAP12	6q24-q25	0.90	1.50	1.66	up	-3.3	-6.1	-5.4	P	210517_s_at
hypothetical gene	LOC389372	6p22.1	0.46	2.99	2.21	proximal	NC/A	NC/A	NC/A	A	1568826_at
hypothetical protein	FLJ10159	6q21	0.78	2.25	1.46	up	NC/A	NC/A	NC/A	A	1563906_at
formin binding protein 1	***FNBP1***	9q34	0.47	1.18	1.09	up	NC/P	-2.7	-3	P	230389_at
homeo box A9	HOXA9	7p15-p14	-0.02	2.14	1.09	up	5.8	6.3	4.6	A	209905_at
zinc finger protein 312-like	LOC389549	7q31.32	ND	1.78	1.09	up	ND	ND	ND	ND	NA
v-maf musculoaponeurotic fibrosarcoma oncogene B	***MAFB***	20q11.2-q13.1	0.40	1.38	1.12	up	-10.9	-10.6	-6.1	P	218559_s_at
hypothetical protein	MGC33530	7p11.2	0.99	1.06	1.84	proximal	NC/A	NC/A	NC/A	A	1554530_at
netrin 4	NTN4	12q22-q23	0.91	2.82	2.43	up	NC/A	NC/A	NC/A	A	234202_at
orthopedia homolog (Drosophila)	OTP	5q13.3	0.89	1.45	1.40	down	NC/A	NC/A	NC/A	A	237906_at
protocadherin 19	PCDH19	Xq13.3	0.45	1.45	1.51	down	NC/A	NC/A	NC/A	A	227282_at
protein tyrosine phosphatase, receptor type, K	PTPRK	6q22.2-23.1	0.80	1.52	1.75	down	NC/A	NC/A	NC/A	A	233770_at
toll-IL 1 receptor (TIR) domain containing adaptor protein	TIRAP	11q24.2	0.83	1.12	1.46	up	NC/P	NC/P	NC/P	P	1552360_a_at
zinc finger protein 37 homolog (mouse)	ZFP37	9q32	0.91	1.71	1.24	proximal	NC/A	NC/A	NC/A	A	207068_at
zinc finger protein 229	ZNF229	19q13.31	0.23	2.64	1.37	proximal	NC/P	NC/A	NC/A	P	1562789_at
zinc finger protein 312	ZNF312	3p14.2	0.22	2.02	1.05	up	NC/A	NC/A	NC/A	A	221086_s_at
zinc finger protein 629	ZNF629	16p11.2	0.98	1.27	1.50	down	-1.7	NC/P	1.2	P	213196_at
deleted in colorectal carcinoma	*DCC*	18q21.3	0.71	1.30	0.34	proximal	NC/A	NC/A	NC/A	A	206939_at
deleted in lung and esophageal cancer 1	*DLEC1*	3p22-p21.3	0.91	1.32	-0.12	proximal	NC/A	NC/A	NC/A	A	207896_s_at
distal-less homeo box 3	DLX3	17q21	0.21	1.22	-0.03	up	NC/A	NC/A	NC/A	A	231778_at
dual oxidase 2	DUOX2	15q15.3	0.82	1.97	-0.25	down	NC/A	NC/A	NC/A	A	219727_at
endothelial PAS domain protein 1	EPAS1	2p21-p16	0.21	2.57	-0.46	down	-3.9	1.8	1.6	P	200878_at
EPH receptor A10	EPHA10	1p34.3	0.65	1.15	0.54	up	NC/A	NC/A	NC/A	A	243717_at
ES cell expressed Ras	ERAS	Xp11.23	0.99	1.00	0.62	up	ND	ND	ND	ND	NA
FERM, RhoGEF and pleckstrin domain protein 1	***FARP1***	13q32.2	0.44	2.57	0.44	proximal/up	NC/A	NC/A	NC/A	A	227996_at
hypothetical protein	FLJ42461	17p13.2	0.45	2.18	0.40	proximal	NC/A	NC/A	NC/A	A	229730_at
gamma-glutamyl hydrolase	GGH	8q12.3	0.28	3.10	-0.10	up	4.4	-2.9	2.6	P	203560_at
glycoprotein V (platelet)	GP5	3q29	0.08	1.36	-0.18	down	NC/P	NC/P	NC/P	P	207926_at
hyperpolarization activated cyclic nucleotide-gated K+4	HCN4	15q24-q25	0.68	1.28	0.53	up	NC/P	NC/A	NC/A	A	206946_at
histone 1, H4I	HIST1H4L	6p22-p21.3	-0.63	3.21	ND	up	NC/A	NC/A	NC/A	A	214562_at
v-jun sarcoma virus 17 oncogene homolog (avian)	***JUN***	1p32-p31	0.11	2.07	-0.08	up	-4.5	-5.1	-4.3	P	201466_s_at
potassium channel beta 3 chain	KCNAB1	3q26.1	0.66	1.48	0.98	down/down	NC/A	5.4	NC/A	A	210471_s_at
protocadherin 8	PCDH8	13q14.3-q21.1	0.79	1.66	-0.09	up	NC/P	NC/P	NC/P	A	206935_at
RAB38, member RAS oncogene family	RAB38	11q14	0.53	1.37	0.55	down	NC/P	NC/P	NC/P	A	234666_at
RAB3C, member RAS oncogene family	***RAB3C***	5q13	0.00	1.57	0.19	proximal	NC/A	NC/A	1.7	A	242328_at
ribonuclease P/MRP 30kDa subunit	RPP30	10q23.31	-0.22	1.68	-0.14	up	0.4	-1	NC/P	P	203436_at
secretogranin III	SCG3	15q21	0.96	2.30	0.12	down	NC/A	NC/A	NC/A	A	219196_at
sonic hedgehog homolog (Drosophila)	SHH	7q36	0.99	1.29	0.50	down	NC/A	NC/A	NC/M	A	207586_at
synaptojanin 2	SYNJ2	6q25.3	0.01	1.15	0.07	up	2.3	2.2	1.3	P	212828_at
zinc finger protein 36, C3H type-like 1	***ZFP36L1***	14q22-q24	-0.02	2.07	-0.03	down	-1.2	-5.5	-3.8	P	211965_at
zinc finger protein 222	ZNF222	19q13.2	0.06	1.56	0.00	proximal	2.2	NC/A	3.4	A	206175_x_at
zinc finger protein 516	***ZNF516***	18q23	-0.51	1.86	-0.12	up	NC/P	-1.6	NC/P	P	203604_at
zinc finger protein 582	ZNF582	19q13.43	0.10	1.27	0.47	proximal	NC/P	NC/P	NC/P	A	1553221_at
zinc finger protein 610	ZNF610	19q13.41	0.96	1.72	0.47	up/down	1.9	NC/A	NC/P	A	235953_at
hypothetical gene	AK055761	12q24.2	0.43	0.75	2.18	down	ND	ND	ND	ND	NA
doublesex and mab-3 related transcription factor 2	***DMRT2***	9p24.3	0.03	0.12	2.08	up	NC/A	NC/A	2.9	A	223704_s_at
NK6 transcription factor related, locus 1 (Drosophila)	NKX6-1	4q21.2-q22	-0.09	0.86	2.18	down	NC/A	NC/A	NC/A	A	221366_at

Hybridization results of CpG island microarrays are presented as mean log2 ratios between normal and tumor cell lines of three independent microarray experiments (log2 ratios above 1 are boxed in black). Results of expression array analysis are presented as mean log2 ratios between normal and tumor cell lines if a significant change was detected (negative log2 ratios indicate lower expression in tumor cell lines and are boxed in black). P, present; A, absent; NC, no change (boxed in gray); ND, not detected. Genes that were independently analyzed by MCIp and real-time PCR (see Figure 5-9 and Figure 5-10) are indicated in bold lettering. Confirmed targets of aberrant hypermethylation are indicated in bold italics. Genes that have been shown to be hypermethylated in other types of tumors are in italics.

5.2.1 Experimental validation of microarray results

A representative number of gene fragments that were identified using combined MCIp-on-chip analyses were selected for further validation. *Mse* I-digested DNA from three different myeloid cell lines (KG-1, THP-1, U937) was subjected to MCIp. Afterwards LightCycler real-time PCR was chosen to validate the MCIp enrichment in the 1000 mM fraction of 29 candidate genes. Out of these, no significant differences were detected at the *LDLR*, *TGIF* and *CBX6* gene fragments. However, 26 gene fragments were enriched in a manner comparable to the results obtained by microarray analyses. The results for the 26 gene fragment validations are represented in Figure 5-9.

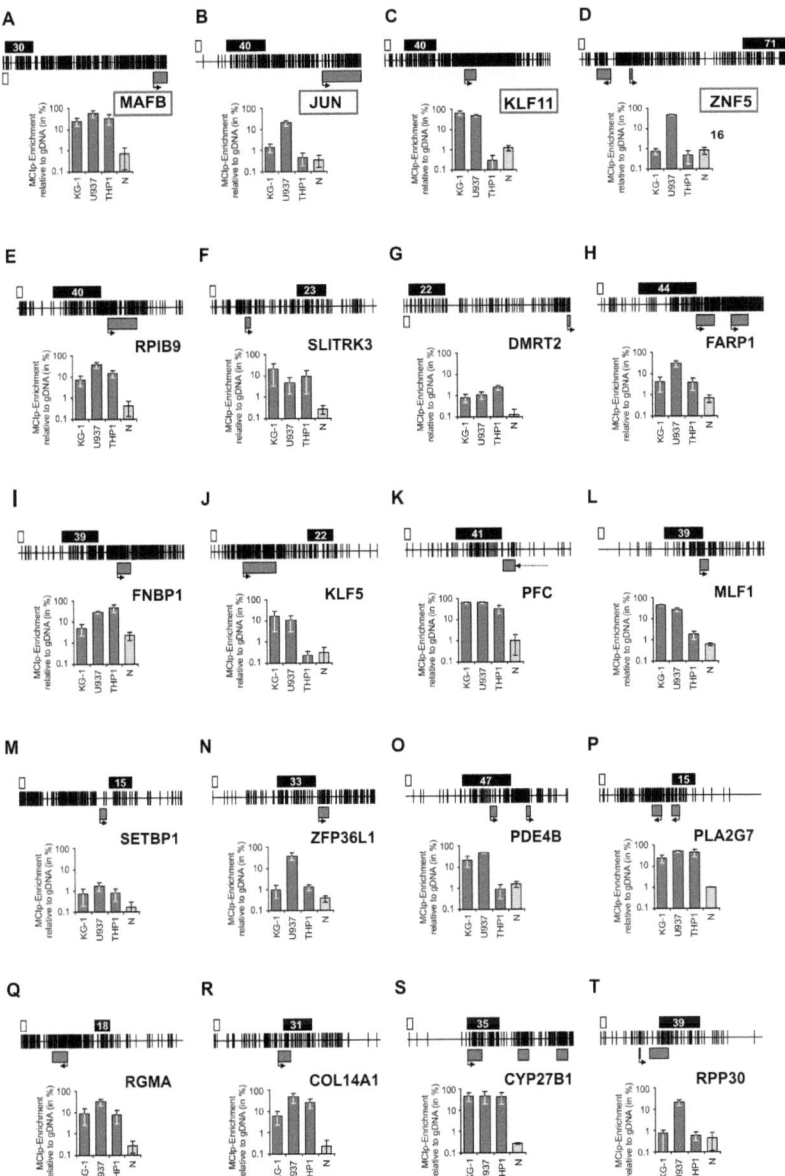

Figure 5-9 Validation of CpG island microarray results by MCIp and real-time PCR
Detection of the methylation status of the indicated human genes in three *Mse* I-restricted human myeloid cell lines (KG-1, U937, THP-1) as well as normal human blood monocytes (N) after MCIp. The corresponding CpG island is represented on top of each figure as described in Figure 5-3. Results are shown relative to the PCR product generated from the input DNA (100%) of each cell type. Values are mean ± SD (n≥3) using at least two different preparations of MBD-Fc. Genes framed in red were validated by bisulfite sequencing (Figure 5-11).

Results

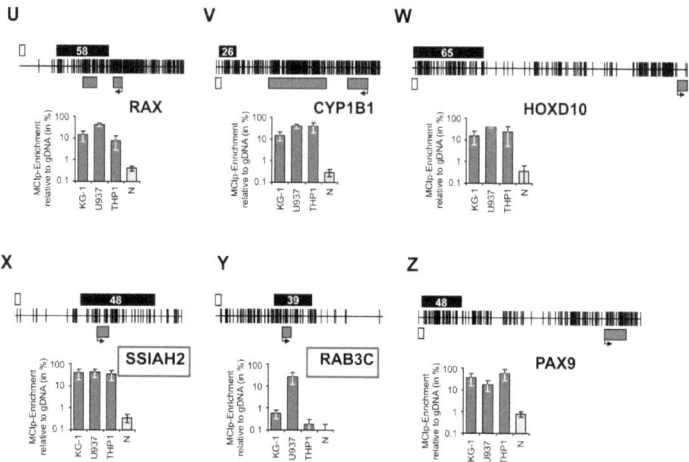

Figure 5-9 continued from page- 95 -

In several cases, the *Mse* I fragment represented on the microarrays did not include the proximal promoter. Since transcription factors may play an important role in leukemogenesis, DNA fragments that include transcriptional start sites of the transcription factor genes *JUN*, *MAFB*, *KLF11* and *ZNF516* were additionally analyzed using MCIp and real-time PCR. While *JUN* promoter fragments were not significantly detected in any of the samples (data not shown), *MAFB*, *KLF11* and *ZNF516* promoter fragments also showed significant methylation (Figure 5-10).

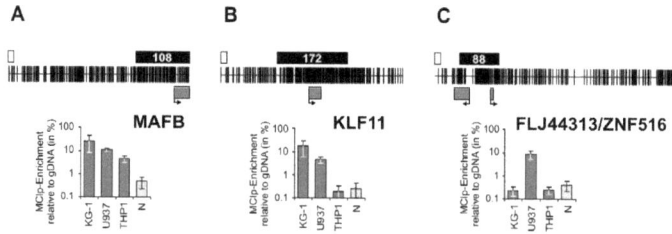

Figure 5-10 Real-time PCR of DNA fragments including transcription start sites
Schematic representation of the MCIp enrichment detected by single gene real-time LightCycler PCR for *MAFB*, *KLF11* and *ZNF516 Mse /Csp6* I promoter fragments in the three leukemia cell lines (*KG-1*, *THP-1* and *U937*) as well as normal human blood monocytes (N). The corresponding CpG island is represented on top of each figure as described in Figure 5-3. Results are shown relative to the PCR product generated from the input DNA (100%) of each cell type. Values are mean ± SD (n≥3) using at least two different preparations of MBD-Fc.

To validate the MCIp detected methylation differences using an independent approach, the methylation status of six CpG island fragments (*JUN*, *RAB3C*, *MAFB*, *KLF11*, *ZNF516* and

Results

SSIAH2 (*LOC283514*)) was additionally analyzed using bisulfite sequencing. As shown in Figure 5-11 the degree of methylation as determined by bisulfite sequencing correlated well with the results obtained by MCIp and real-time PCR validation.

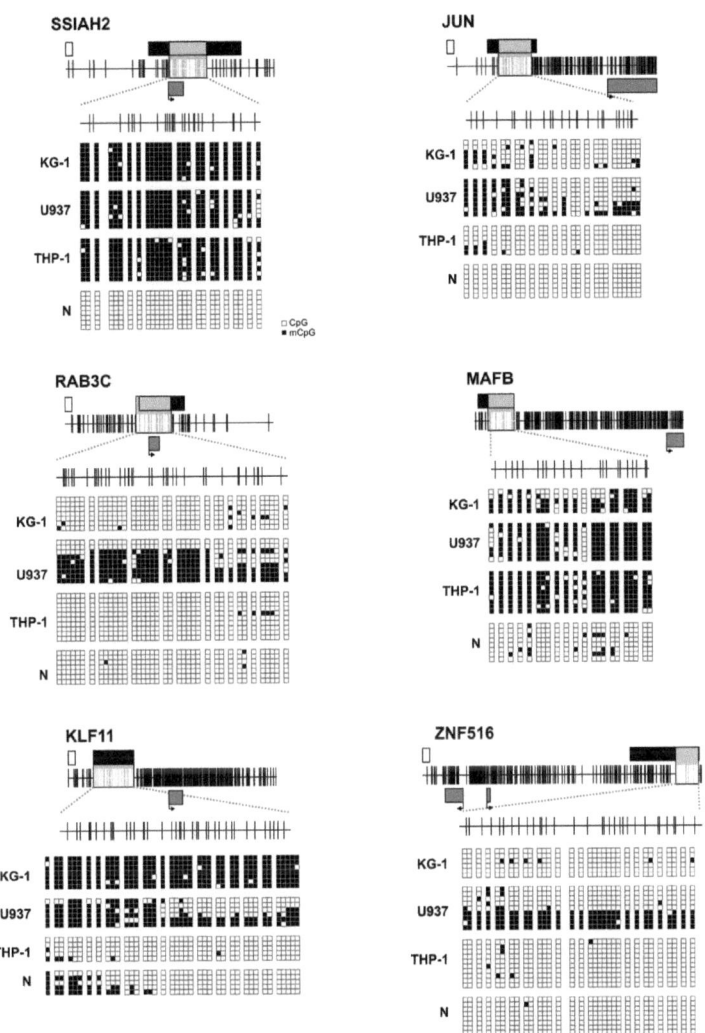

Figure 5-11 Bisulfite sequencing of six differentially methylated gene loci
DNA methylation was analyzed in genomic DNA from three human myeloid leukemia cell lines (KG-1, THP-1, U937) as well as normal blood monocytes (N) at the represented loci by bisulfite sequencing. The CpG islands are represented on top of each figure as described in Figure 5-3. The gray boxes represent the regions that were amplified from bisulfite-treated DNA and cloned. Several independent inserts were sequenced and results are presented schematically. Squares mark the position of CpG dinucleotides (empty: unmethylated; filled: methylated).

5.2.2 Global comparison of CpG island methylation and mRNA expression

RNA from KG-1, U937, and THP-1 cells as well as from freshly isolated human blood monocytes of a healthy donor was isolated and complementary mRNA expression data were generated by microarray experiments using the human HGU-133 Plus 2 Array from Affymetrix. Table 5-1 represents a side by side comparison of CpG methylation and mRNA expression concerning the identified CpG island fragments. Interestingly, more than half of the genes (69/125) were undetectable in all samples using the microarray approach. In cases where significant mRNA levels were detected, transcription was often downregulated in tumor cell lines compared to normal monocytes when the gene was methylated. Examples include *PLA2G7*, *FNBP1*, *MAFB*, or *ZNF516*. In some cases there was no correlation between the degree of methylation and gene expression (e.g. *HOXA10*, *EPAS1*, or *SYNJ2*). In those cases CpG methylation probably targets regions not relevant for enhanced transcription.

To confirm the downregulation of a few representative genes with hypermethylated CpG islands (*JUN*, *MAFB*, *KLF11*, *SSIAH2*, and *ZNF516*) in leukemia cell lines compared to human blood monocytes, reverse transcription and quantitative real-time PCR analysis were performed (data not shown). A significant derepression in U937 cells could be achieved when treated with 5 µM decitabine (5-aza-2'-deoxycytidine). Figure 5-12 demonstrates the effect of demethylation which was most striking for *MAFB* and *SSIAH2* that were induced up to 100-fold in treated cells.

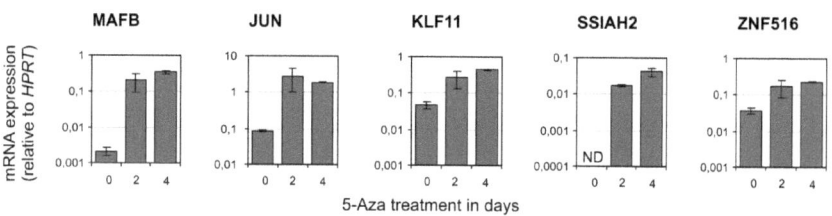

Figure 5-12 Derepression of hypermethylated target genes by decitabine
Quantitative real-time PCR of *MAFB*, *JUN*, *KLF11*, *SSIAH2*, and *ZNF516* detects the mRNA expression levels in myeloid leukemia cell lines at the time point 0 and after 2 and 4 days of decitabine treatment (5 µM). Expression levels are relative to *HPRT* expression. Value are mean ± SE of two experiments. ND, not detected.

5.2.3 Aberrant hypermethylation in patients with acute myeloid leukemia

Tumor cell lines only represent *in vitro* models of primary tumors. Cell lines often have acquired additional alterations both on genetic and epigenetic levels. It has been reported that a large proportion of genes are hypermethylated across multiple cancer cell lines, suggesting that these differences are due to intrinsic properties in generating cell lines (Smiraglia et al., 2001). The potential role of culture effects has been further highlighted by a recent study demonstrating that DNA methylation profiles of human embryonic stem cells vary over time in culture, with different genes affected in different cell lines (Allegrucci and Young, 2007).

To test whether genes that were found to be hypermethylated in the leukemia cell lines are also affected in primary tumors, DNA from blast cells derived from twelve AML patients was analyzed for hypermethylation at 21 different promoter loci (Figure 5-13). For each locus a number of patients showed significant hypermethylation compared to normal donors. In the case of the *PFC* gene, nine AML patients were markedly hypermethylated whereas at the *RAB3C* or *RPIB9* loci at least two patients showed significant enrichment. These results suggest that the CpG island fragments identified in tumor cell lines can also be subject to hypermethylation in primary tumor cells and may represent novel disease markers for leukemias. It is well established that regional methylation levels tend to increase with age in the mammalian genome. Notably, the youngest patient (20 years old, P20) was hypermethylated at 15 (out of 21) loci, whereas the eldest patient (67 years old, P07) was significantly hypermethylated only at the *PFC* locus, indicating that methylation of the above tested loci does not correlate with aging.

Results

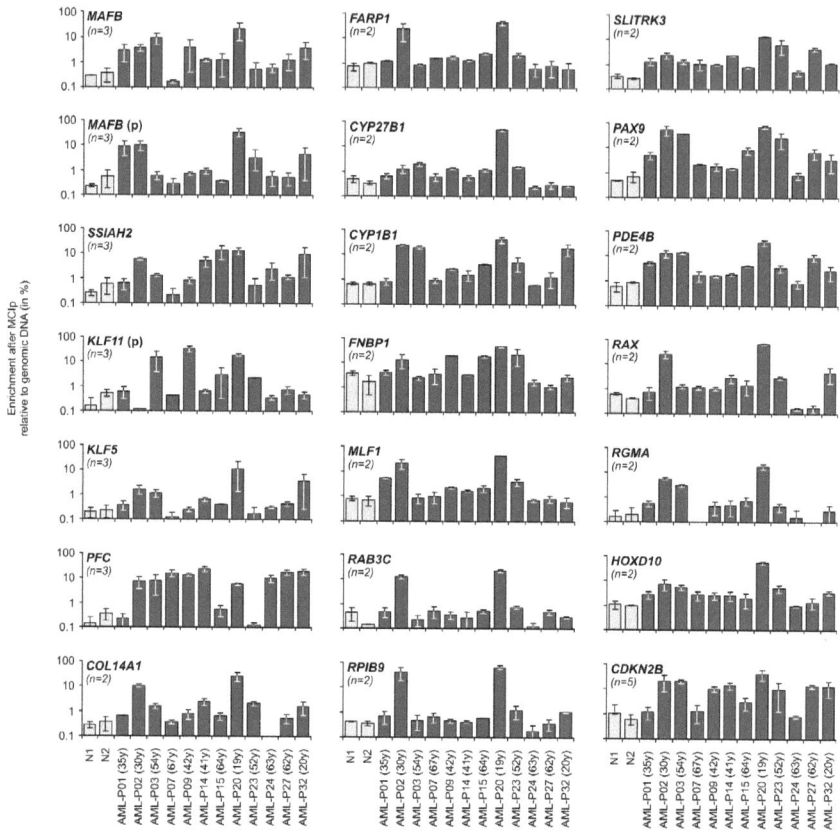

Figure 5-13 Methylation profiles of AML patients
Schematic representation of the MCIp enrichment detected by single real-time LightCycler PCR for the 21 indicated promoter fragments in Mse I/Csp6 I-restricted DNA in the twelve AML samples (AML-Patient number. (age in years)) as well as two normal human blood monocytes (N1, N2). Results are shown relative to the amount of PCR product generated from the input DNA (100%) of each cell type. Values are mean ± SE of at least two LightCycler amplifications using at least two different preparations of MBD-Fc. p means that the fragment is located within the promoter region and n is the number of experiments performed.

5.3 Global profiling of cancer-associated CpG island hypermethylation using MCIp combined to 244K CpG island arrays

5.3.1 Establishment of a new microarray platform

In our early studies the MCIp technique in combination with human 12K CpG island microarrays (HCGI12K, Microarray Center, UHN, Toronto, Canada) was used to identify more than one hundred genes with aberrantly methylated CpG islands in three myeloid leukemia cell lines. These results showed that the MCIp technique discriminates DNA fragments according to the methylation degree and allows an unbiased genome-wide detection of hypermethylation.

However, the initial experiments performed with the Human CpG 12K microarrays highlighted several issues. Besides quality problems, many other aspects encouraged us to switch to another microarray platform. The previously used 12K microarray platform contained many genomic *Mse* I fragments with high variation in fragment length. In addition, the array contained repetitive fragments leading to unwanted cross-hybridization events (non-specific binding), which possibly gave rise to misleading results. Furthermore, the number of the representative genes on the array was relatively small. Thus, for global analysis of patient samples, another array platform provided by the company Agilent seemed to be better suited for this purpose. This array contains 244,000 probes (50-60 mer oligonucleotides) covering about 23,000 CpG islands within coding and non-coding regions of the human genome (Agilent 244K CpG island microarrays).

To adapt the fractionation approach to the new Agilent DNA microarrays, several modifications were required. Instead of the previously used *Mse* I digestion, genomic DNA was sonicated to a mean fragment size of 350-400 bp. Sonication of genomic DNA leads to a statistical fragmentation which is necessary for an unbiased genome-wide methylation profiling. Moreover, large-scale MCIp (4 µg DNA instead of 300 ng) was used in order to provide sufficient amount of gDNA for subsequent labeling and microarray hybridization. Therefore PCR bias caused by ligation-mediated amplification (LM-PCR) (see section 4.3.6.2) could be avoided. Empirical evidence showed that the coating of the protein A sepharose beads influences the fractionation behavior. In order to compensate for possible variations concerning the coating (due to varying quality of different protein

Results

batches) prior to each set of fractionation of the normal and tumor samples, a test MCIp with DNA derived from the U937 cell line was performed with a part of the freshly coated beads to define the cut-off for highly methylated DNA. The individual MCIp fractions from U937 were spin-purified (PCR purification kit, Qiagen) and eluted in 100 µl EB buffer. Subsequently the fractionation of U937 DNA was controlled by qPCR using control primers covering the imprinted region of *SNRPN* as well as a genomic region lacking CpGs (Empty 6.2) and the CpG island region *RPIB9* (strongly methylated in U937). While both alleles of the imprinted *SNRPN* are eluted in different fractions (the unmethylated one with a low salt buffer, the methylated one with a high salt concentration), the bulk of the unmethylated CpG empty region (negative control) is enriched in the low salt fractions due to the complete absence of CpGs (data not shown). In contrast, the *RPIB9* fragments which are highly methylated in the U937 cell line were detected in the high salt fractions. According to these results a threshold is defined at a salt concentration which separates the strongly methylated DNA fragments from the intermediate and low methylation fragments. Assuming that the other samples were enriched for methylated DNA in the same manner, MCIp was performed with the actual samples (tumor and normal samples) according to the determined cut-off. The high salt fractions containing the highly methylated CpG island fragments were directly labeled for microarray hybridization. Cancer cell DNAs were labeled with Alexa Fluor 647 and DNA from normal cells was labeled with Alexa Fluor 555 using the Bioprime Plus Array CGH Genomic labeling System (Invitrogen, Carlsbad, CA, USA). Efficiency of the labeling reaction was controlled with UV-spectroscopy and comparative hybridization on CpG island oligonucleotide microarray was performed using the recommended protocol (Agilent). Image data was extracted with Agilent feature extraction software and imported to Microsoft Excel for further analysis.

Because the signal intensities were quite low compared to the background noise and the method was not as robust as expected, the application had to be further improved. Different conditions were tested to achieve optimal results. First, after MCIp, each fraction was purified using MinElute Columns to reduce the volumes (elution in 20 µl EB). Therefore loss of DNA by lyophilization could be circumvented. To achieve a better control after MCIp, not only a pretest with U937 DNA was performed, but all individual samples were controlled by qPCR with control primers to determine the cut-off. Moreover to attain improved labeling, a new Kit from Invitrogen was used (Bioprime total Genomic Labeling System). Consequently, enriched methylated DNA fragments of the high salt MCIp fractions were labeled with Alexa Fluor 5-dCTP (cancer cells) and Alexa Fluor 3-dCTP (normal cells).

Since GC-rich probes have the tendency to cross-hybridize, the stringency of hybridization was increased by a combination of a higher incubation temperature (67°C instead of 65°C) and by addition of formamide (15%) to the hybridization reaction mix. Therefore misleading results such as false positive and false negative signals should be minimized. This step probably had the major impact on better results. Figure 5-14 demonstrates that using stringent hybridization conditions, many more probes could be detected as hypermethylated in the tumor cell line.

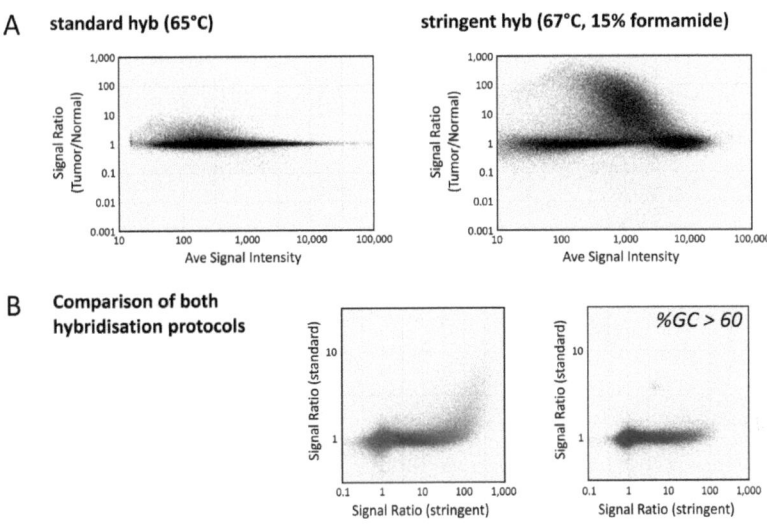

Figure 5-14 Comparison of both hybridization protocols
(A) The two diagrams demonstrate the difference between the standard (65°C, without formamide) and the stringent (67°C, 15% formamide) hybridization protocol. The signal ratio between U937 and monocytes is shown as a function of the average signal intensity. (B) The diagrams demonstrate the difference when the signal ratios of both protocols are directly compared showing that with the stringent protocol much more probes are detected as hypermethylated than with the standard hybridization conditions, especially when the probes have a high GC content (>60%).

To explore if the signal intensities increased with greater quantities of DNA, MCIp and subsequent microarray analysis were performed with 1 μg, 2 μg and 4 μg DNA. Figure 5-15 and Figure 5-16 compare data using different amounts of DNA and different hybridization protocols. Figure 5-15 illustrates a comparison of genome-wide

Results

hypermethylation profiles, whereas Figure 5-16 depicts three selected regions (*FOXP3*, *MARVELD2*, *IRX3*). The studies showed that robust methylation profiles could be obtained with as little as 1 µg of genomic DNA using the stringent (new) protocol. Best results were achieved with 4 µg of genomic DNA while 2 µg DNA were sufficient for good, reproducible results.

Because we were limited in patient material, all following experiments were performed with 2 µg DNA instead of 4 µg DNA.

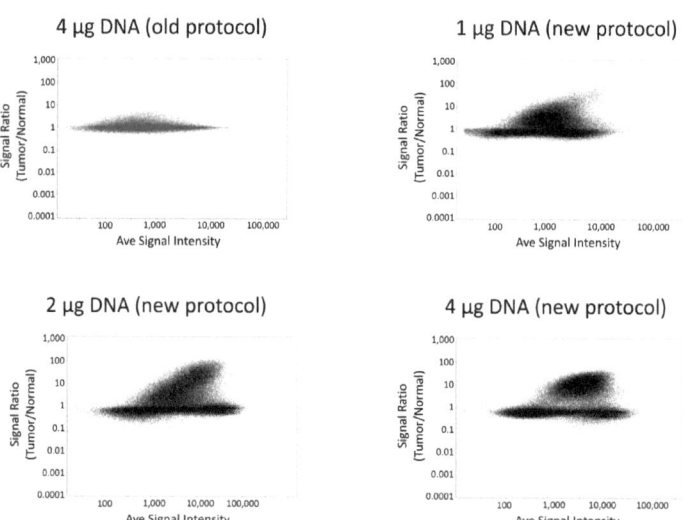

Figure 5-15 Major modifications of the MCIp-on-chip protocol in global screening for tumor-specific hypermethylation
The old hybridization protocol using standard hybridization conditions (65 °C and no formamide) was compared to the newer stringent hybridization protocol (67 °C and 15% formamide). The new stringent protocol was performed with three different amounts of input DNA (1 µg, 2 µg, 4 µg subjected to MCIp). The signal ratios between tumor and normal DNA were plotted as a function of the average signal intensity.

Results

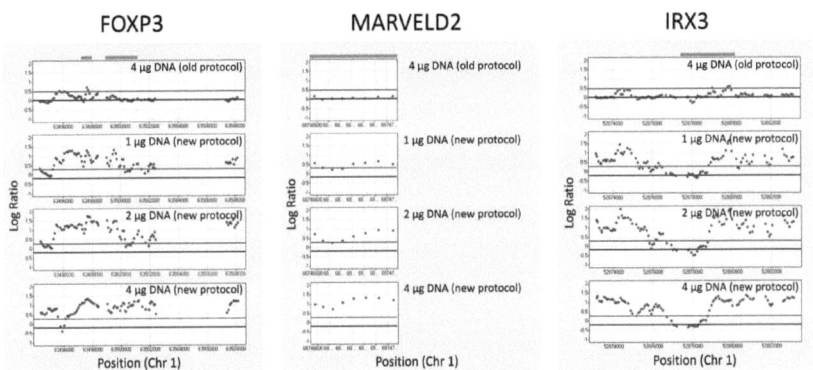

Figure 5-16 Examples of microarray results using different hybridization conditions and increasing amounts of DNA
Shown are data points for three CpG island regions of *FOXP3*, *MARVELD2* and *IRX3* using different hybridization conditions. Each data point represents one microarray probe. The old standard protocol involved hybridization at a temperature of 65°C, whereas in the new stringent protocol the hybridization temperature was increased to 67°C and 15% formamide was added. The log ratio (tumor/normal) is plotted as a function of the relative position on chromosome 1.

5.3.2 Comprehensive validation of genome-wide CpG island methylation profiles for two human leukemia cell lines

Data from this section have been published in the journal *Cancer Research*. Comparative MCIp hybridization data were deposited with the GEO Data Library under Series Entry: GSE17455, GSE17510, GSE17512.

To establish and test the newly adapted and improved MCIp-on-chip technique, the first genome-wide methylation analyses using Agilent 244K CpG island microarrays were performed with MCIp-enriched methylated fragments of the well-established leukemia cell lines U937 and THP-1 in comparison with enriched fragments of blood monocytes of a healthy donor.

Because the signal intensities were biased in correlation to their GC content (higher GC content lowered the average signals), the probe signals were GC normalized. Three independent replicates of each cell line were highly similar (mean r^2=0.79 and 0.87 for $\log_{10}$ ratios of THP-1- and U937-monocyte comparisons, respectively). A typical scatter plot (Figure 5-17) of a comparative hybridization of MCIp-enriched material from U937 cells highlights the three types of hybridization behavior: probes that show low signal intensities in both samples (absence of DNA methylation), probes indicating specific

Results

enrichment (aberrant DNA methylation) in the leukemia samples, and probes that show high signal intensities but low signal ratios in both samples (methylated in both samples). Altogether, more than one third of all microarray probes showed significant enrichment in the U937 cell line.

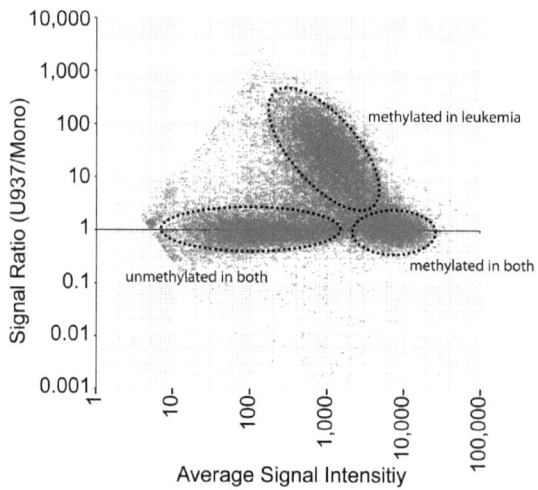

Figure 5-17 Comparative DNA methylation analysis of U937 cells and normal human blood monocytes using methyl-CpG immunoprecipitation (MCIp)
Representative scatter plot of a comparison of MCIp-enriched material from 2 µg genomic DNA of a leukemia cell line (U937) and a pool of normal human blood monocytes from healthy donors on human 244K CpG island arrays. Signal intensity ratios are plotted against average signals (MvA plot, $\log_{10}$ scale). More than one third of all microarray probes show significant enrichment in the cell line.

All CpG islands that were validated as hypermethylated in the first study (with 12K CpG island arrays) were again detected as hypermethylated in these experiments (performed with 244K Agilent arrays). In total, approximately 11,300 or 8,700 (out of 23,000) independent regions were significantly enriched or depleted (>2.5-fold different) in U937 or THP-1, respectively. The majority of differentially enriched regions showed signs of hypermethylation (or amplification) in both cell lines. In U937 cells 10,700, in THP-1 cells 6,800 differential methylated regions (DMR) were detected. Hypomethylation or deletion of individual regions were mainly found in THP-1 cells.

Large-scale validation of MCIp microarray data using mass spectrometry analysis was performed to quantify methylation differences at the resolution of single CpGs. Therefore a representative set of differential methylated regions (DMRs) as well as regions without methylation difference between cell lines and monocytes as a negative control were chosen to be analyzed by the MassARRAY system. This method is based on Matrix-Assisted Laser Desorption/Ionization Time-Of-Flight Mass Spectrometry (MALDI-TOF MS) measurement of bisulfite converted DNA using the EPITYPER platform (Sequenom, San Diego, US). Bisulfite treatment generates methylation dependent sequence variations, which can be measured by the MassARRAY system. Moreover, this procedure allows the analysis of multiple CpGs in one amplicon and the comparison of their methylation status between different samples (for more details see section 4.4.6). The validation panel comprised a set of 140 genes that were selected based on the comparative MCIp methylation profiles of 23,000 CpG islands from the two myeloid leukemia cell lines (U937 and THP-1). In total, 1,150 primer pairs, which cover about 13,500 CpG sites, were designed for the amplification of bisulfite-treated DNA. Besides THP-1 and U937, MALDI-TOF MS was also performed with DNA derived from monocytes of three different healthy donors and with unmethylated and fully methylated DNAs as controls. The complete MALDI-TOF MS data are provided in the supplementary part of the corresponding publication (Gebhard et al., 2010).

Validation using mass spectrometry analysis of bisulfite-treated DNA (MassARRAY System Sequenom) was highly consistent with the microarray data. Figure 5-18 shows examples of microarray-MassARRAY comparisons for four genes (*MLL*, *SMAD6*, *HOXB5* and *EPAS1*) which demonstrate a high degree of correlation between both approaches.

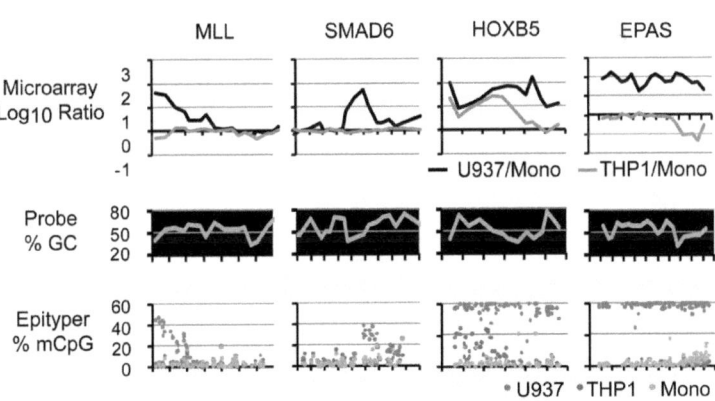

Figure 5-18 Examples for correlation between MCIp and bisulfite data
The panel on top shows microarray results ($\log_{10}$ ratios; black: U937 versus monocytes; gray: THP-1 versus monocytes) for selected CpG island regions (*MLL*, Chr 8: 9798000-97992000; *SMAD6*, Chr 15: 64781200-64783000; *HOXB5*, Chr 17: 44025400-44026600; *EPAS1*, Chr 2: 46378600-46380800). The middle panel depicts the GC-content of microarray probes. The bottom panel shows quantitative MALDI-TOF MS (EpiTYPER) methylation levels for individual CpG residues within the analyzed region. Each spot represents one CpG unit.

Results

In a next step all results extracted from both data sets were compared. To correlate differential signal intensities on CpG island microarrays with data points for individual CpG dinucleotides, the mean methylation difference between tumor cell lines and monocytes of all measured CpG dinucleotides that are located in a 300 bp radius around a microarray probe was calculated ("EpiTYPER methylation ratio", for further information see section 4.4.6.10). Figure 5-19 illustrates the high consistency of both approaches regarding the comparison of the two leukemia cell lines THP-1 and U937.

When plotting the "methylation ratio" against the probes' signal ratio, a good correlation ($r^2 = 0.51$ for U937 and $r^2 = 0.67$ for THP-1) was observed between microarray probe intensity ratios on microarrays and mean bisulfite methylation ratios of CpG dinucleotides located around the microarray probe for the U937 as well as the THP-1 cell line. The high reproducibility of both approaches is shown exemplary for U937 leukemia cells and normal monocytes in Figure 5-20A.

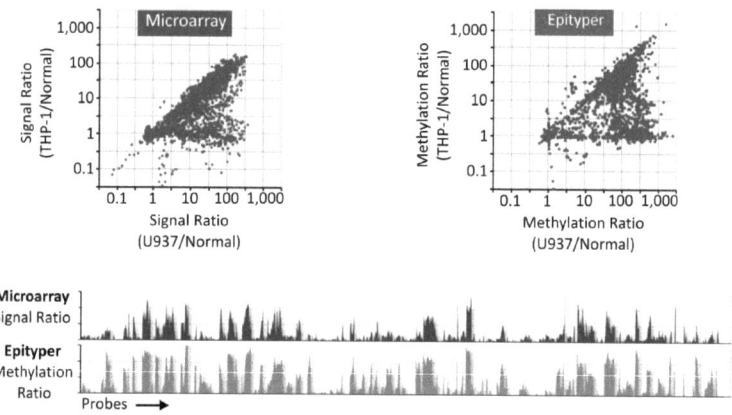

Figure 5-19 Correlation of microarray and mass spectrometry data
Differential methylation between the two cell lines THP-1 and U937 can be reliably detected using both methods, MCIp combined to microarray analysis and MALDI-TOF MS. Differential methylation of the microarray data is illustrated by plotting the signal ratios from one cell line against the signal ratios from the other cell line. For the EpiTYPER data methylation ratios for both cell lines are calculated and plotted against each other. In the lower graph, the high consistency between microarray signal ratios and EpiTYPER methylation ratios is illustrated on probe level.

A similar comparison was performed to correlate both data sets based on region level instead of probe level. Here, mean methylation ratios ($\log_{10}$) for 225 regions (covering more than 300 bp each) (see section 5.4.2) based on the above amplicons were calculated and then plotted against mean signal ratios of all microarray probes within each region. As shown in Figure 5-20C and D, a good correlation between microarray and EpiTYPER data was observed ($r^2 = 0.62$ for U937 and $r^2 = 0.75$ for THP-1).

The correlation between both methods increased on the region level compared to the probe level, mainly because the resolution of the microarray approach dropped at extremely GC-rich microarray probes which tended to cross-hybridize, even under stringent hybridization conditions. Mass spectrometry data was also able to provide higher resolution of the boundaries between methylated and unmethylated domains. Stretches of unmethylated DNA in close vicinity to metylated domains were often detected as methylated due to the DNA fragmentation range. Nevertheless, the MCIp approach clearly discriminated methylation levels between the two leukemia cell lines U937 and THP-1 (Figure 5-20B and Figure 5-19). Thus, our comprehensive validation demonstrates a good overlap of MCIp and MALDI-TOF MS data and suggests that our technique allows for reproducible and valid detection of comparative CpG methylation levels.

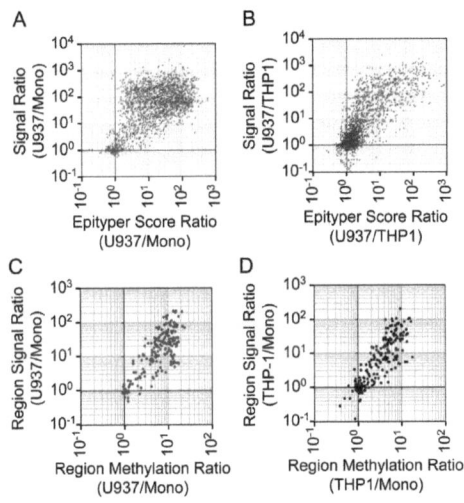

Figure 5-20 Methyl-CpG immunoprecipitation and its validation using MALDI-TOF MS
(A) Microarray probe-based correlation of MALDI-TOF MS (EpiTYPER) and MCIp microarray results of U937/monocyte comparisons. Microarray probe signal log_{10} ratios were plotted against an EpiTYPER score that consists of a scaled, average methylation level of all CpGs located in a radius of 300 bp around the microarray probe ($r^2 = 0.51$ for U937 and $r^2 = 0.67$ for THP-1, data not shown). (B) Microarray probe-based correlation of differential CpG methylation ratios measured by MALDI-TOF MS (EpiTYPER) and MCIp microarray. MCIp reliably detects differential methylation between the two cell lines ($r^2 = 0.65$). Correlation of MALDI-TOF MS (EpiTYPER) and MCIp microarray results of U937/monocyte (C) and THP-1/monocyte (D) comparisons for regions covered by the EpiTYPER analysis. Mean probe signal log_{10} ratios are plotted against mean log_{10} transformed EpiTYPER methylation ratios ($r^2 = 0.62$ for U937 and $r^2 = 0.75$ for THP-1).

5.3.3 Genome-wide hypermethylation profiling in AML and patients with colorectal carcinoma

Cancer is associated with disease-related epigenetic abnormalities, including the aberrant hypermethylation of CpG islands which can lead to the abnormal silencing of tumor suppressor genes. A major challenge of current clinical research is to find ways of exploiting the diagnostic and therapeutic implications of these abnormalities. Since hypermethylation of CpG islands seems to be a tumor-type specific event, the knowledge of global DNA methylation patterns of a given tumor might provide important information for risk assessment, diagnosis, monitoring and treatment. The design for a study to screen patient samples to find out new potential marker genes is shown in Figure 5-21.

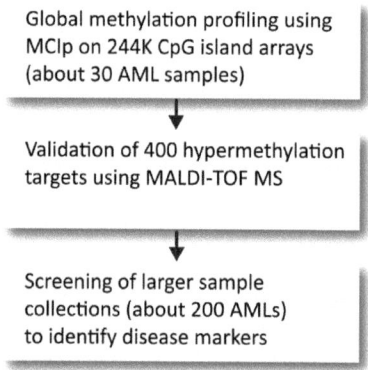

Figure 5-21 Study design for identifying disease markers for AML
In principle, the study comprises three steps that should finally help to identify disease markers.

Since global methylation profiling of tumor cell lines showed high sensitivity and reproducibility of the MCIp approach using the stringent (new) protocol, in a next step global comparative CpG island methylation profiling for more than 25 AML samples with mostly normal karyotype was performed, using MCIp in combination with 244K CpG island microarrays. As a reference a pool of DNA derived from three different healthy donors (male) was used. Image-data was extracted using Feature Extraction Software 9.5.1 (Agilent) and the standard CGH protocol. Processed signal intensities were further normalized using GC-dependent regression and imported into Microsoft Office Excel 2007 for further analysis. (Complete microarray data sets will be submitted with the corresponding publication.)

Clustering of samples based exclusively on the X- and Y-chromosomal genes demonstrated that male and female samples can be clearly distinguished as expected due to an enrichment of the X-chromosomal gene fragments and simultaneous depletion of the Y-chromosomal genes (Figure 5-22). This resulted from the fact that the reference pool consisted of DNA derived from three different healthy male donors.

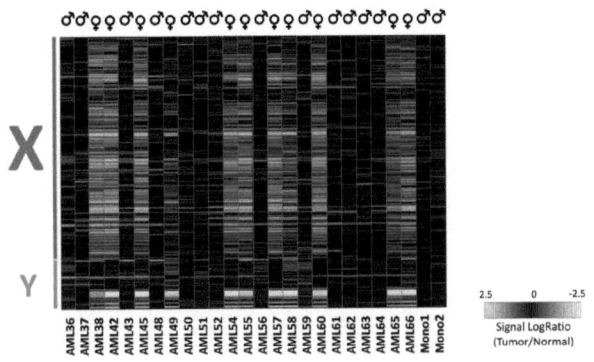

Figure 5-22 Hierarchical cluster analysis of AML samples in X- and Y-chromosomal genes only
DNA methylation ratios (tumor/normal) from 244K CpG island microarray analyses are represented on a continuous scale from non-methylated (yellow) to fully methylated (blue). Male (♂) and female (♀) samples can be clearly distinguished due to an enrichment of the X chromosome gene fragments and depletion of the Y chromosome gene fragments.

To evaluate differences in the methylation patterns between 27 AML patients and three leukemia cell lines (KG-1, U937, THP-1) as well as normal blood monocytes and colon cells from a healthy donor, a hierarchical cluster analysis was performed (Figure 5-23). The analysis was limited to autosomal genes in order to account for the effects introduced from X-chromosomal imprinting.

Results

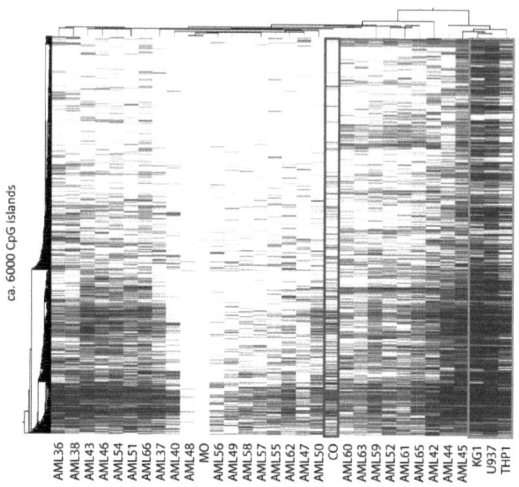

Figure 5-23 Hierarchical clustering of tumor samples and one monocyte as well as one colon sample
Two-way hierarchical cluster analysis of 27 AML samples, 1 colon sample from a healthy donor (CO), 1 monocyte sample from a healthy donor (MO) and 3 leukemia cell lines (KG-1, U937, THP-1) are shown in the different columns. CpG island regions are represented as lines on a continuous scale from non-methylated (white) to fully methylated (black).

Hierarchical clustering revealed a complex methylation pattern of AML patients. The results demonstrate that more than 6,000 CpG island regions (out of 23,000) were hypermethylated in at least three AML patients. At this level there was no obvious correlation between karyotype, age and other parameters. Cell lines showed a much higher degree of methylation than primary tumors. About 3,000 CpG islands were methylated in cell lines but never in primary tumor cells. Furthermore, the methylation pattern of the normal colon tissue DNA showed that many CpG islands methylated in AML become also methylated in the colon during aging. This became apparent by comparing hypermethylation profiles of healthy colon samples (derived from 60-year-old donors) and monocyte samples (derived from 20- to 30-year-old healthy donors). Colon samples showed a high degree of hypermethylation compared to monocytes: 6,000 out of 23,000 CpG island regions can be hypermethylated in AML samples, 3,000 (out of 6,000 CpG island regions) were also hypermethylated in normal colon samples. This high amount of hypermethylated regions cannot only be reduced to tissue-specific effects. They rather represent age-dependent differences. Genes affected by hypermethylation during aging are mainly genes involved in developmental processes like homeobox genes or Polycomb targets as demonstrated in Figure 5-24.

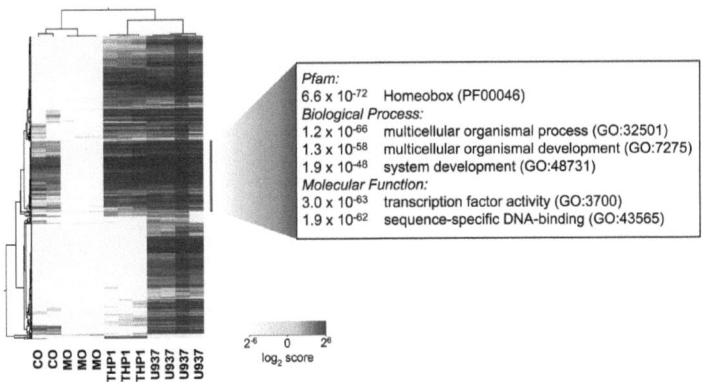

Figure 5-24 Age-related hypermethylation correlates with developmental genes
Hierarchical clustering revealed a group of 3,000 CpG islands that were methylated not only in U937 and THP-1 cell lines but also in normal colon samples, however not in normal blood monocytes. Genes associated with this cluster are enriched for gene ontology (GO) terms including homeobox genes, developmental processes and transcription factors. Three independent experiments for THP-1 and four independent experiments for U937 showed high consistency, as well as two independent experiments performed with different normal colon samples (CO) and three independent experiments performed with normal blood monocytes derived from different healthy donors (MO).

One possibility to make sure that those differences are really age-dependent, would be the comparison with colon samples from a set of younger healthy donors. However, during the time of my thesis there were no samples available for such experiments.

Because a large number of CpG islands showed variable methylation aberrations in AML, we decided to focus on genes that are likely to include functionally relevant candidates. In total, 400 target regions that are important for transcription, gene regulation, or signaling, were chosen for validation using the MassARRAY platform (MALDI-TOF MS). Although we include some examples for putative age-dependent methylation, the majority of amplicons did not fall into this category.

So far, we concentrated on leukemia samples. To compare the methylation profiles obtained from AML samples with an independent tumor entity, about 20 DNA samples derived from colorectal carcinomas were analyzed using MCIp with subsequent microarray hybridizations. A reference pool of three different colon DNAs (50-, 56- and 63-year-old donors) purchased from Lonza were also used. Complete microarray data sets will be submitted with the corresponding publication. The data analysis of the colon patients has not been completed at time of writing this thesis. Future aims are the mass spectrometric analysis and validation of hypermethylated genes to define potential marker genes for this tumor entity.

5.3.4 Confirmation by MassARRAY (EpiTYPER) data

Mass spectrometry yields quantitative methylation data of short stretches of subsequent CpGs in a high-throughput manner and consequently allows for the validation of large genomic regions. Approximately 350 genes that were differentially methylated between AML patients and normal monocytes were selected based on the array results (Figure 5-23). For all genes a total of 670 PCR amplicons were designed. Before analyzing the patient samples, all primers were tested with the cell lines (THP-1 and U937). The results of all amplicons were screened for selection for follow-through methylation analysis in the AML samples. Several criteria were considered when choosing the best amplicons:

1. Spectral quality: All amplicons designed across a CGI were assessed for spectral quality from the MALDI-TOF MS output to determine the success of PCR amplification and the presence / absence of primer dimers or amplification bias.

2. Cleavage pattern: The cleavage pattern of each amplicon following base-specific cleavage was also assessed to determine whether a sufficient amount of cleavage fragments fell within the mass range of detection for MALDI-TOF MS (1,500 Da - 6,500 Da). In some situations, a large proportion of cleaved fragments were either too small or too large for detection.

3. CpG density and length: Larger (>400 bp), CpG-dense amplicons were preferable, in order to maximize the quantity of data available.

4. Location: Amplicons adjacent to or upstream of the transcription start site were considered ideal, in order to cover any putative transcription factor binding sites.

5. Methylation levels: When methylation ratios were considered across several amplicons covering a CGI, amplicons in regions where methylation levels changed dramatically between samples (from unmethylated to methylated or vice versa) or where tumor sampes were methylated were preferable, rather than amplicons where no methylation was observed.

Following manual inspection of all methylation data quality, a final set of about 400 amplicons (>7000 CpG sites) in about 300 genes were chosen for validation of the microarray data with AML patient samples.

Again, MassARRAY EpiTYPER data correlated well with microarray data. Examples for the excellent consistency of the two different techniques are shown in Figure 5-25.

We finally asked, if specific markers for disease diagnosis or prognosis can be identified. To address this issue, a further 200 AML patients were screened using the 400 amplicons as described above to identify relevant disease markers. (The complete MALDI-TOF MS data will be available online upon publication.)

Figure 5-26 shows exemplary methylation patterns of two different gene amplicons (*CEBPA* and *RHOB*) for 165 AML patients, CD34+ cells derived from three different healthy donors and monocytes derived from four healthy 20-year-old donors and eleven healthy 60-year-old donors. *CEBPA* (CCAAT/enhancer binding protein α) is a basic leucine zipper transcription factor that regulates differentiation-dependent genes during granulocyte differentiation. While hypermethylation of the *CEBPA* promoter has already been reported in AML as well as in other malignancies (Figueroa et al., 2009), the distal CpG island which is located about 20 kb downstream of the promoter was often hypomethylated in AMLs but methylated in normal monocytes and stem cells (Figure 5-26). In contrast, more than one third of the analyzed AML patients showed significant hypermethylation of an amplicon within the *RHOB* gene whereas hematopoietic stem cells (CD34+ cells) as well as monocytes of all healthy donors were unmethylated. It is already known that the expression of the *RHOB* gene, a member of the *Rho* family of small GTPases, is often downregulated in lung cancer (Sato et al., 2007).

Computational analyses of the EpiTYPER data set was still in progress at the time of writing this thesis. Therefore, no correlations between methylation profiles and clinical parameters could be detected or, likewise, no potential marker genes could be identified at this time.

Results

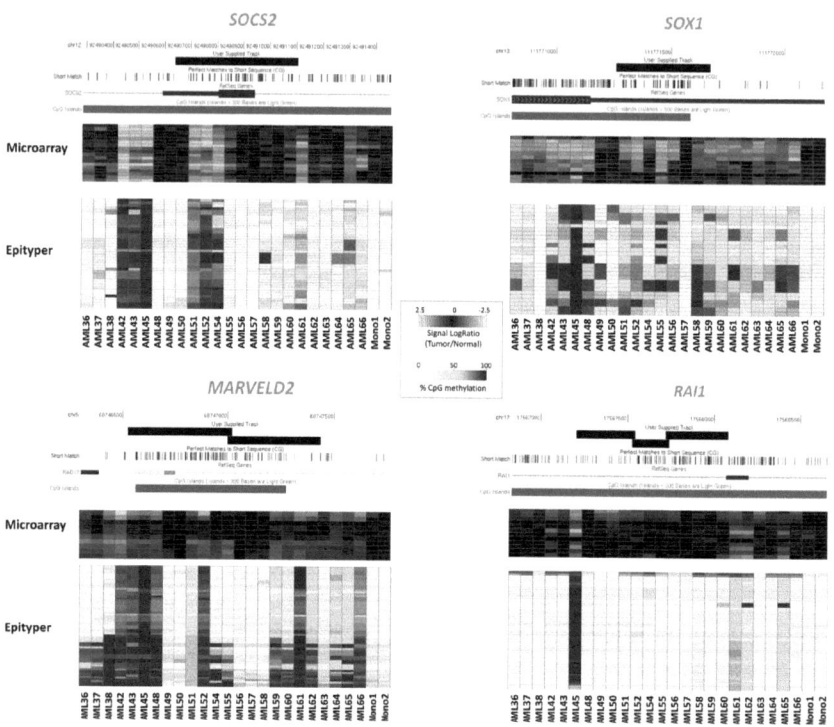

Figure 5-25 Examples of aberrantly methylated CpG islands in AML samples
Microarray and MassARRAY data are shown for CpG islands of four different genes in AML samples and two blood monocytes from two different healthy donors. Each sample is represented by one column. Each line of the microarray results represents one probe and each line of the EpiTYPER results represents one CpG unit. The same region is detected by microarray or EpiTYPER analysis, respectively. DNA methylation values regarding EpiTYPER results are represented on a continuous scale from non-methylated (white) to fully methylated (dark blue) (non-detectable CpGs are marked in gray), whereas signal log ratios (tumor versus normal) are represented on a continuous scale from blue (strongly hypermethylated in tumor) to yellow (strongly hypomethylated in tumor). The top diagrams were extracted from the Genome Browser showing the relative position of transcripts, CpG islands (green) as well as position of amplicons detected by MassARRAY experiments.

To find out if monocytes also show age-related differences in methylation patterns similar to colon samples, and therefore to make sure that the identified potential marker genes are really methylated due to tumorigenesis and not due to aging, DNA samples derived from monocytes of about 60-year-old donors were analyzed using MALDI-TOF MS for the 400 CpG island regions. Unlike colon samples, monocyte samples did not show age-dependent changes in DNA methylation (Figure 5-26). One explanation could be that crypt stem cells possess an exceptionally high rate of proliferation, resulting in further DNA methylation due to the higher mitosis rate.

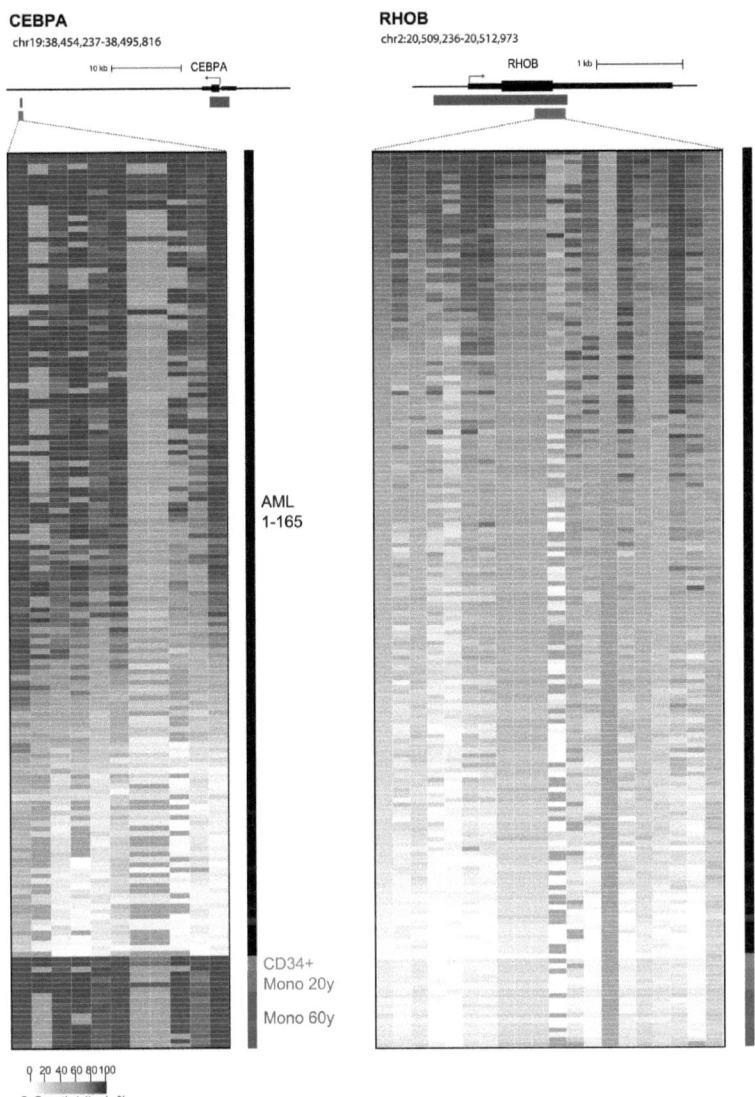

Figure 5-26 Examples of abnormal methylation patterns in AML patients
Mass spectrometry analysis of CpG island fragments of two different genes (*CEBPA* (on the left) and *RHOB* (on the right)) in 165 AML patients, three CD34+ samples and monocytes derived from four healthy 20-year-old donors and eleven healthy 60-year-old donors. Samples (rows) are clustered according to the average methylation degree of all CpG units (columns) within the amplicon. DNA methylation values are depicted by a color scale as indicated (methylation increases from white (non-methylated) to dark blue (fully methylated)). Gray denotes data of poor quality. The top diagrams were extracted from the Genome Browser showing the relative position of transcripts (black), the transcription start sites (arrows), CpG islands (green) and the position of amplicons detected by MassARRAY experiments (red).

5.4 General transcription factor binding at CpG islands in normal cells correlates with resistance to *de novo* methylation in cancer

The data from this section have been published in the journal *Cancer Research*. Microarray data were deposited with GEO (gene expression analyses: GSE16076; comparative MCIp hybridizations: GSE17455, GSE17510, GSE17512; ChIP-on-chip hybridizations: GSE16078).

Cancer is associated with disease-related epigenetic abnormalities, including the aberrant hypermethylation of CpG islands leading to loss of tumor suppressor gene expression. Methylation profiling studies have demonstrated that although there may be hundreds of different CpG islands methylated in any one tumor, some are methylated in multiple tumor types, whereas others are methylated in a tumor-type specific manner. Moreover, each tumor type tends to exhibit a characteristic set of aberrantly methylated genes. However, despite numerous examples of methylation–associated gene silencing events in human cancer cells, the molecular pathways underlying aberrant DNA methylation remain elusive. Different mechanisms for cancer-dependent, aberrant *de novo* methylation have been proposed so far, largely based on the behavior of individual CpG islands (see section 1.7.2). Besides other proposed mechanisms, one possible mechanism suggests that *Alu* and other repetitive elements may serve as foci from which *de novo* methylation can spread (Feltus et al., 2003), whereas other elements could provide a "protective" function. The absence of such "protective" transcription factors may lead to the spreading of DNA methylation into affected CpG islands (Turker, 2002). To address this issue, methylation-prone and methylation-resistant CpG islands were defined by analyzing the methylation status of 23,000 CpG islands of the human genome in acute leukemia cell lines as well as normal blood monocytes. Understanding the nature of these differences could provide insight into the molecular basis for aberrant methylation. The corresponding MCIp-on-chip experiments as well as the extensive validation by bisulfite conversion and subsequent MALDI-TOF are described in 5.3.2.

Results

5.4.1 Basic properties of hypermethylated CpG islands

Analyzing the results obtained from three comparative microarray hybridizations from monocytes and two leukemia cell lines, about 9,000 CpG island regions were detected reproducibly different (p<0.05 for repeated measures, three replicates each) between any of the comparisons (monocyte - monocyte (MO), THP-1 - monocyte (THP-1), U937 - monocytes (U937)). Hierarchical clustering of these CpG island regions revealed a group of CpG islands that are commonly hypermethylated in the cell lines but not in normal monocytes. Genes associated with this cluster exhibited an enrichment for gene ontology terms related to developmental processes, transcription factor or receptor functions, as well as homeobox proteins, that are often targeted by Polycomb group repressors. These associations were highly significant and in line with earlier evidence suggesting a link between aberrant DNA methylation in cancer and Polycomb group repressors that often target genes involved in development like homeobox transcription factors (Bracken et al., 2006b). CpG islands that were specifically methylated only in one of two leukemia cell lines were not significantly enriched for functional properties (data not shown). The heatmap is represented in Figure 5-27A.

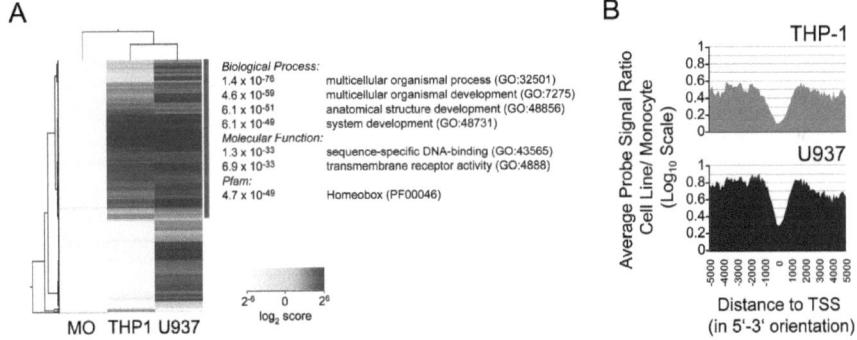

Figure 5-27 Functional analysis of commonly hypermethylated CpG island regions
(A) About 9,000 CpG island regions were significantly different (p<0.05 for repeated measures) between any of the comparisons (monocyte - monocyte (MO), THP-1 - monocytes (THP-1), and U937 - monocytes (U937)). Hierarchical clustering of these CpG island regions reveals a group of CpG islands that are commonly hypermethylated in the cell lines. Genes associated with this cluster are enriched for gene ontology (GO) terms including developmental processes, transcription factors and homeobox genes. (B) Averaged microarray signal intensities were plotted as a function of distance towards the known transcription start sites (TSS; bin size 100 bp, motif in 5'-3' orientation).

Plotting the average probe signal ratio between one of the two cell lines (THP-1 and U937) and normal blood monocytes as a function of the distance to the transcription start site (TSS) reflects that regions around known TSSs are less often targeted by *de novo* methylation in leukemia cells than promoter distal sites (Figure 5-27B). This is confirmed by previous observations suggesting that proximal promoters are less frequently *de novo* methylated than other genomic regions (Irizarry et al., 2009).

5.4.2 Defining CpG island regions

CpG methylation often spreads over large genomic regions. In order to perform a region-based analysis of comparative methylation data instead of a microarray probe-based analysis, regions were first mapped and a cumulative hypermethylation value based on probe behavior within the region was assigned. Based on the array design, CpG island regions were defined to include all neighboring microarray probes of a region with a maximum distance of 500 bp and a minimum of three microarray probes in total. For each of the approximately 23,000 CpG island regions an integral value for hypermethylation based on area size was calculated and log_{10} intensity ratios of microarray probe signals were smoothed above a threshold (2.5-fold enrichment) as described in Figure 5-28A. To normalize for region size, these integral values were divided by one hundredth of an arbitrary 'maximal' integral value that was calculated for each region assuming a 100-fold enrichment of each microarray probe in leukemia versus normal DNA.

Figure 5-28B shows a diagram of the cell line U937, where normalized integral hypermethylation ratios are plotted against the corresponding average signal intensities of individual areas. The comparative analysis on the region level (Figure 5-28B) also clearly separated the three different classes of CpG island regions as shown before on probe level (Figure 5-17): (i) CpG islands that show low signal intensities in both samples. Those were lost during the MCIp procedure and are thus unmethylated. (ii) CpG islands that are specifically enriched and therefore methylated specifically in the leukemia samples. (iii) CpG islands that show high signal intensities in both samples. They were enriched in both samples and are thus methylated in both. We could not identify specific properties associated with the latter type of CpG islands which is heterogeneous and includes monoallelic (e.g. imprinted regions) as well as biallelic (tissue- or soma-specific) DNA methylation events (Straussman et al., 2009) and therefore concentrated on properties of unmethylated or *de novo* methylated genes in the following analyses.

Results

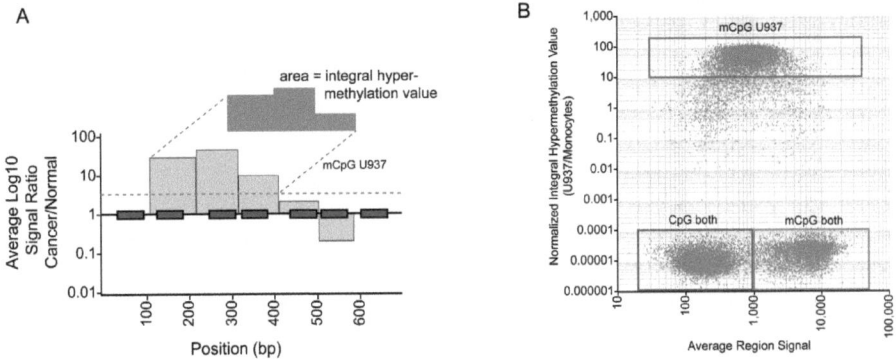

Figure 5-28 Integral hypermethylation values and DNA methylation status in CpG island regions
(A) Calculation of integral hypermethylation values for a CpG island region. Probes (black boxes) within a region (except the ones located at the edges) were assigned mean $\log_{10}$ ratios of the center probe and the two neighboring probes. The integral hypermethylation value assigned to each CpG island region corresponds to the area on the top and is the sum of integral probe areas (light gray boxes) above the threshold: $\sum$[extended probe length x (smoothed $\log_{10}$ ratio-2.5)] All probes with a smoothed $\log_{10}$ ratio below the threshold (2.5, dotted line) were ignored.
(B) Normalized integral hypermethylation values of CpG island regions are plotted against their average log signal intensities for a comparative MCIp of U937 cells and monocytes. Three populations are distinguished: CpG islands that show low signal intensities in both samples (low average log signal intensities and low integral hypermethylation values: lost during the MCIp procedure and thus unmethylated; CpG both); CpG islands that are characterized by intermediate average log signal intensities and high integral hypermethylation values (specifically enriched and methylated in the leukemia samples; mCpG U937); CpG islands that show high signal intensities (but low signal ratios) in both samples (high average log signal intensities and low integral hypermethylation values: enriched in both samples and thus methylated in both; mCpG both).

To get insight into the correlation of DNA methylation and mRNA expression, the previously performed CpG island microarray hybridizations (U937-monocyte, THP-1-monocyte) were compared with global mRNA expression analyses, which were performed with RNA isolated from monocytes (CD14+ cells), CD34+ progenitor cells and the leukemia cell line U937.

Confirming earlier observations (Gebhard et al., 2006b; Keshet et al., 2006), the comparison of global mRNA expression data of normal (CD14+ cells and CD34+ cells) and leukemia cells (U937) between the two major CpG island classes (unmethylated in cancer as well as normal cells and *de novo* methylated in cancer cells) demonstrated that the majority of *de novo* methylated CpG islands are characterized by low or absent transcription of neighboring genes irrespective of their position relative to TSS (promoters, intragenic or intergenic regions) in normal as well as in cancer cells. Box plots in Figure 5-29 illustrate the distribution of mRNA expression ratios.

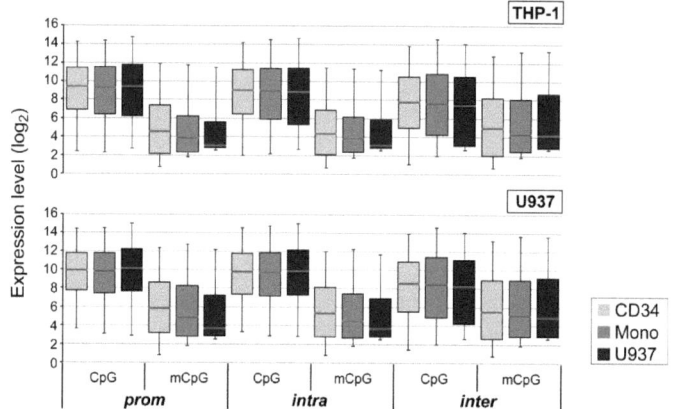

Figure 5-29 Expression status of genes associated with CpG island regions
The box plots show the distribution of mRNA expression ratios (CD34+ progenitor cells, CD14+ normal blood monocytes, U937) conditional on the methylation status (unmethylated in both: CpG; hypermethylated in leukemia: mCpG) for THP-1 and U937 at individual probes that were divided into the three position classes: promoters (prom), intragenic (intra) and intergenic (inter). The red lines denote medians, boxes the interquartile ranges, and whiskers the 5th and 95th percentiles. Pair wise comparisons of mRNA expression ratios associated with unmethylated and hypermethylated regions are significant (P<0.001, Mann–Whitney U test, two-sided).

5.4.3 Strategies for *de novo* motif discovery

If certain transcription factors are to be involved in establishing and/or maintaining the CpG methylation status of a certain CpG island region, it should be possible to isolate their respective binding motifs by comparing methylation-prone and methylation-protected CpG islands. A number of previous bioinformatics attempts indeed identified specific nucleotide sequences and general CpG island attributes (Das et al., 2006; Feltus et al., 2003; Feltus et al., 2006) or structural features (Bock et al., 2006) that contribute to the protection from or susceptibility to aberrant methylation. However, no defined consensus sites for known transcription factors could be identified so far. After analyzing the methylation status of 23,000 CpG islands of the human genome we defined sequence patterns characteristic for methylation states in CpG island regions using *de novo* motif analysis.

Motif discovery was performed using the comparative algorithm HOMER (**H**ypergeometric **O**ptimization of **M**otif **E**n**R**ichment) (see section 4.5.1 for more details). Different strategies for the analysis of the data in this thesis were applied in order to avoid identifying biased results due to CpG island nucleotide content, length etc. CpG island regions intrinsically have a high CpG content (in contrast to a large fraction of the genome that is relatively CpG-poor) and they differ in sequence lengths. *De novo* motif finding, however, requires target and background sequence sets of constant lengths and similar nucleotide compositions.

The following approaches were used to isolate enriched motifs: We first searched for motifs in all selected CpG island regions (unmethylated in both or methylated only in tumor) in a fixed radius around the CpG island region center (± 250, 500, 750, and 1000 bp) against either a non-overlapping, CpG-matched background set from all known promoters, or from all microarray-defined CpG island regions. Independently, CpG island regions were divided into regions <750 bp, 750-1250 bp, 1250-1750 bp and >1750 bp and separately analyzed as above. These region sizes were chosen because the CpG islands in this study ranged in size from 0.2-3.2 kb and because *cis*-acting features might lie within or flanking CpG islands. The identified motifs as described below as well as slight modifications thereof generally appeared in most analyses. The region approach is always associated with some degree of impreciseness, since some CpG island regions (in particular the larger ones) tend to display heterogeneous methylation patterns. For example, if a CpG island is methylated on one side, but unmethylated on the other side, the whole region will be considered as methylated. When compared with our large bisulfite data sets (see section 5.3.2), approximately 10% (U937) or 20% (THP-1) of the CpG island regions' overall hypermethylation score did not match the methylation status of the actual sequence around the motif (data not shown).

Because the resolution of region data is relatively low, we applied a more accurate motif centered approach to improve resolution and accuracy of motif-methylation status correlations. For this purpose each of the identified motifs on the CpG island array was annotated with the mean signal intensities of all microarray probes in the range of ± 150 bp around it. The lower limit for hypermethylation of a motif in the tumor cell line was set at the mean signal intensity $\log_{10}$ ratio of 0.8. A motif was counted as unmethylated if it had the mean signal intensity $\log_{10}$ ratio between 0.4 and -0.4 and an average $\log_{10}$ signal intensity below 3.0-3.5 depending on the overall microarray probe behavior in individual hybridizations. In comparison with the bisulfite data, this approach was much more accurate: in U937, 135/140 motifs (97%) and in THP-1 124/135 motifs (92%), were correctly classified as hypermethylated or unmethylated.

5.4.4 Sequence motifs associate with CpG island regions that remain unmethylated or become hypermethylated in cancer

The hypothesis that CpG islands differ in their inherent susceptibility to aberrant methylation presupposes that there are *cis*-acting features that distinguish methylation-prone and methylation-resistant CpG islands. To address this question, the *de novo* motif discovery algorithm HOMER was used to search for sequence patterns associated with CpG island regions that are either specifically and highly methylated in leukemia cell lines or not methylated in any sample. Altogether a set of eight non-redundant sequence motifs could be identified that were highly enriched in either population in comparison with all CpG island regions on the array (Figure 5-30A). These motifs were highly similar to known matrices from the TRANSFAC database. Hypergeometric P values for the enrichment of the indicated sequence motifs were assigned based on motif-centered methylation data. The calculation was performed with the mean signal intensities of all microarray probes in the range of $\pm$ 150 bp around each motif (see section 5.4.3). Two repetitive motifs were highly enriched in the hypermethylated CpG island set, one of them (GAGA) (P value = 5.7×10^{-51} for U937) resembling the consensus motif for *Drosophila* GAGA-binding factor, a trithorax group member that has been implicated in preventing heterochromatin spreading. CA-repeats (CACA) (P value = 7.5×10^{-76} for U937) may play a role in RNA splicing and are bound by the heterogeneous nuclear ribonucleoprotein (hnRNP) L in a repeat length dependent manner. But there is no known link to DNA methylation or chromatin structure. More strikingly, the *de novo* motif algorithm revealed six sequences highly enriched in the unmethylated CpG island population. Five of them corresponded to consensus binding sites for known transcription factors, including nuclear transcription factor (NF) Y, GA binding protein (GABP), specific protein (SP) 1, nuclear respiratory factor (NRF) 1, ying-yang (YY) 1, whereas one of them was an unknown factor. The latter motifs were enriched with high significance with hypergeometric P values from 10^{-148} to $<10^{-300}$ (motif distribution and P values are submitted with the corresponding publication). The ratios of expected versus observed motif appearance show the clear enrichment/depletion of the above motifs in unmethylated or methylated CpG island regions in U937 and THP-1 cells, respectively (Figure 5-30B).

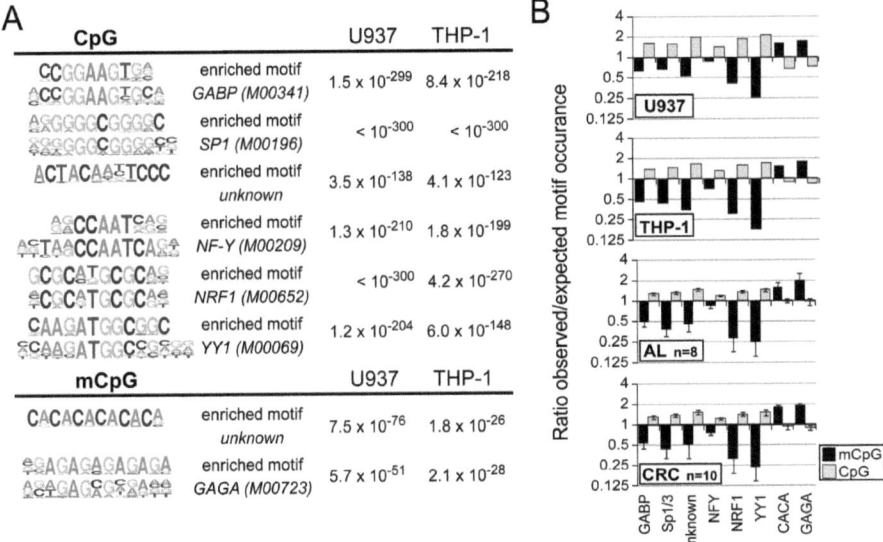

Figure 5-30 Sequence motifs associated with aberrantly DNA methylated (mCpG) and commonly unmethylated CpG island regions (CpG)
(A) P values (hypergeometric) for the enrichment of the indicated sequence motifs were assigned based on motif-centered methylation data (based on mean signal intensities of all microarray probes in the range of ± 150 bp around each motif. Motifs identified *de novo* are shown in comparison to known matrices from the TRANSFAC database. (B) The two upper diagrams depict ratios of observed versus expected motif occurrences in CpG island regions that are aberrantly DNA methylated specifically in cell lines (mCpG) or unmethylated in monocytes and the cell lines (CpG). The distribution of sequence motifs was also analyzed in acute leukemia samples (AML n=8) or colorectal carcinomas (CRC n=10). Here, median ratios of observed versus expected motif occurrences are shown as described above. Error bars mark the interquartile range. Hypergeometric P values for individual enrichments are listed in the supplementary material of the corresponding publication.

Cell lines have been extensively cultured and may therefore have acquired genetic and epigenetic alterations that are not necessarily found in primary cells. To explore whether the sequence motifs identified in the two cell lines (individual motif distribution and P values are available online in the supplementary part of the corresponding publication) were also apparent in primary tumors, comparative methylation profiles of eight samples from acute leukemia (compared to normal monocytes) (see section 5.3.3) were analyzed concerning to the distribution of the above identified motifs. All sequence motifs were again significantly enriched in either unmethylated or methylated CpG island regions in primary AML samples (Figure 5-30B) indicating that the identified motifs are also relevant *in vivo*.

Results

To obtain evidence whether the protective role of the identified motif panel was also relevant in a different class of tumor, we analyzed ten colorectal carcinomas (compared to normal colon) (see section 5.3.3). Again, the same set of motifs showed a high enrichment in either unmethylated or methylated CpG island regions (Figure 5-30B). Thus, the provided data strongly suggest that the identified consensus sequences are of general importance and may serve to protect CpG islands (preferably those acting as promoters) from aberrant methylation.

Next, similar analyses were also performed with groups of unmethylated or methylated CpG island regions that were classified according to their genomic position (promoter: -1000 - +100 of RefGene TSS; intragenic (all exons and introns of RefGenes); intergenic: all non-transcribed regions). All six sequence motifs identified in CpG island regions that are unmethylated in normal cells and also remain unmethylated in tumor cells were enriched within the proximal promoter regions of known genes. Ratios of observed versus expected motif occurrence are demonstrated in Figure 5-31A. This is in line with previously published data that describe motifs isolated from unmethylated CpG islands as prominent constituents of proximal promoters (Rozenberg et al., 2008; Xie et al., 2005b). In Figure 5-31B the distribution of motifs is illustrated with respect to transcription start sites (TSS) of known genes. All motifs except the two repeat sequences are enriched within proximal promoters. Interestingly, most of the discovered promoter motifs (GABP, Sp1, NFY, NRF1 and the unknown motif) show positional bias with respect to TSS towards the 5'-direction.

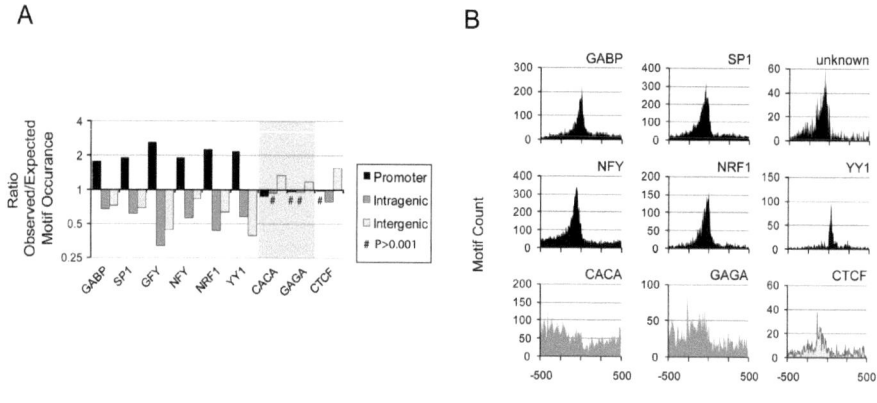

Figure 5-31 Motif enrichment in cell lines depending on genomic location
(A) The distribution of sequence motifs was analyzed conditional on their genomic location. Enrichments or depletions at the three position classes were highly significant (hypergeometric test: P<0.001) except for the cases marked with a hash. (B) Distribution of motifs relative to transcription start sites of known genes (TSS). With the exception of the two repeat sequences, all motifs show strong peaks close to proximal promoters.

Results

In contrast, the two repeat sequences (CACA, GAGA) showed no specific enrichment around transcription start sites (TSS), but both showed a higher enrichment in promoter proximal than in distal sites that acquired methylation during leukemic transformation (data not shown). Motif searches conditional on their genomic position additionally identified a CTCF consensus motif specifically enriched in the unmethylated intergenic CpG island regions (Figure 5-31). The 20 bp motif used for further analysis was extracted from published Chip-sequencing data (Barski et al., 2007) using HOMER:

CTCF motif:

$$\text{ATAGTGCCACCTAGTGGCCA}$$

Despite the significant over-representation of the "protective" motifs in promoters, they were also enriched with high significance in unmethylated CpG island regions that were located in intergenic or intragenic regions as shown in Figure 5-32 for U937 and THP-1 cells. In contrast, CTCF, a transcription factor which can act as a chromatin barrier by preventing the spread of heterochromatin structures, showed only enrichment in the intergenic regions in both cell lines.

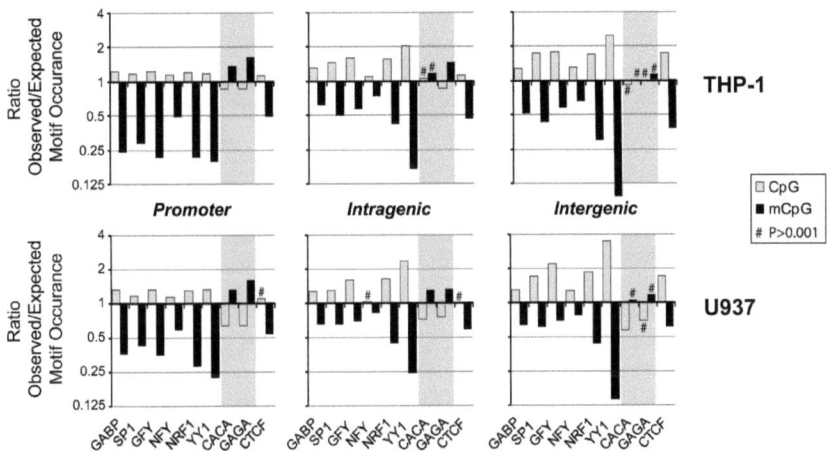

Figure 5-32 Sequence motifs associated with aberrantly methylated (mCpG) and commonly unmethylated CpG island regions (CpG) depending on their genomic location
The diagrams depict ratios of observed versus expected motif occurrences at sequences that are hypermethylated specifically in cell lines (mCpG) or unmethylated in monocytes and cell lines (CpG). Enrichments or depletions at the three position classes were highly significant (hypergeometric test: P<0.001) except for the cases marked with a hash.

If a certain factor was able to confer methylation protection, this property should be limited to its vicinity. Therefore distal sequences should be less protected than proximal ones. By

Results

plotting average MCIp signal intensities (normalized for GC-content) as a function of motif distance for each of the "protective" motifs, it could be demonstrated that each of the protective motifs showed a similar distribution of signal ratios: values were lowest at the center and progressively increased with distance. Curve progression was flat in primary, normal cells. However, distance-related differences in signal ratios markedly increased in leukemia cells, suggesting that these motifs are indeed associated with lower methylation levels and that this association depends on motif distance (Figure 5-33). Interestingly, the signal ratio distribution was not always symmetrical (e.g. at the unknown motif or the NRF1 motif), implying that some factors may preferentially protect regions upstream or downstream of the element. Repeat elements showed an inverse distribution: mean signal ratios were usually higher at the motif center and tended to drop with distance. The distribution of signal ratios also appeared unsymmetrical. This is consistent with the preferential *de novo* methylation of CpG regions located up to 2 kb distant from CpG island promoters. Those so-called CpG island shores were previously detected in colon cancer (Irizarry et al., 2009).

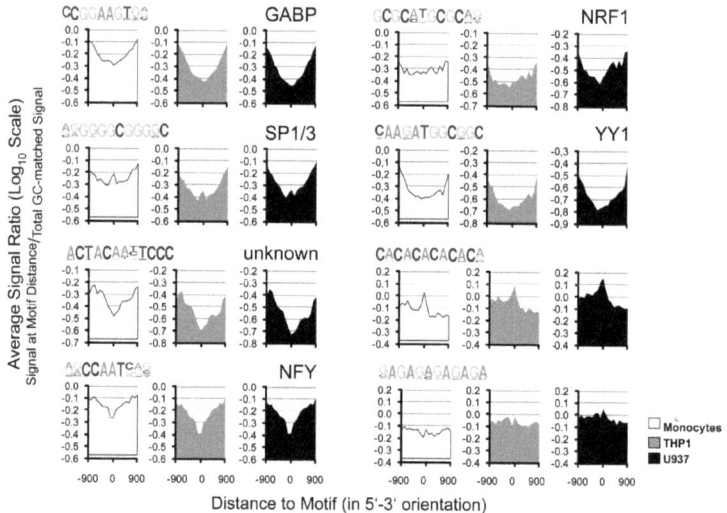

Figure 5-33 Distribution of DNA methylation relative to motif distance in monocytes and leukemia cell lines
Averaged microarray signal intensities were calculated as a function of motif distance (bin size 100 bp, motif in 5'-3' orientation), distance dependent values were normalized for the average, GC-content matched signal intensities of all array microarray probes and ratios were plotted against motif distance. Negative values are below the average of all microarray probes (and therefore less methylated), whereas positive values imply that regions are methylated above the average of all regions.

Results

This phenomenon was not only evident in the human genome but also in the murine genome. Averaged DNA methylation ratios of individual CpGs derived from high-throughput Reduced Representation Bisulfite Sequencing (RRBS) of mouse embryonic stem (ES) cells, ES-derived and primary neural cells, and eight other primary tissues (Meissner et al., 2008) were calculated for all available CpG dinucleotides and plotted as a function of motif distance (bin size 100 bp, motif in 5'-3' orientation). As shown in Figure 5-34 a similar, motif distance-dependent distribution in the masked as well as in the unmasked mouse genome became obvious. Interestingly, on this global (less CpG island biased) scale, the relatively short NFY motif was less protective, especially when annotated on the unmasked mouse genome. A likely explanation may be that the ratio between actual factor binding and motif occurrence is lower in non-CpG island regions.

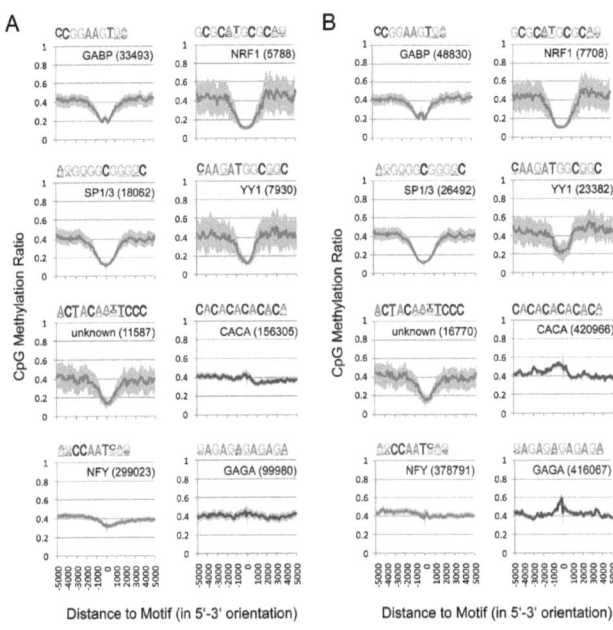

Figure 5-34 Distribution of DNA methylation relative to motif distance in murine ES cells
CpG methylation ratios were extracted from high-throughput Reduced Representation Bisulfite Sequencing (RRBS) data (GEO database accession no. GSE11034) of murine embryonic stem (ES) cells, ES-derived and primary neural cells, and eight other primary tissues. Average CpG methylation ratios were calculated for all available CpG dinucleotides and plotted as a function of motif distance (bin size 100 bp, motif in 5'-3' orientation). Motifs are mapped against a repeat-masked (A) or unmasked (B) mouse genome. On this global (less CpG island biased) scale, the relatively short NFY motif was less "protective", especially when annotated on the non-repeat-masked (unmasked) mouse genome which may be related to the fact that the ratio between actual NFY binding and motif occurrence is lower in non-CpG island regions. Total number of motifs is given in brackets. Gray areas represent distance-dependent standard deviations of CpG methylation ratios.

Results

5.4.5 Sequence motifs and transcription factor binding in normal cells correlate with CpG methylation status in leukemia

To study the correlation between motif appearance, transcription factor binding in normal cells and aberrant DNA methylation in the tumor cell lines, ChIP-on-chip (chromatin immunoprecipitation combined to microarray) analyses with antibodies for the transcription factors Sp1, NRF1 and YY1 in normal peripheral blood monocytes were performed. The distribution of binding events was analyzed based on their genomic location (promoter, intergenic and intragenic regions) (for ChIP-on-chip peak calling and motif annotation see section 4.5.2). As their consensus sites, these three general factors preferentially bound to promoter regions (Figure 5-35A). Enrichments or depletions at the three position classes were highly significant (hypergeometric test: P<0.001). Furthermore they often bound in the vicinity (± 250 bp) of each other as illustrated in Figure 5-35B. Using the bound regions defined by ChIP-on-chip experiments, *de novo* motif analysis revealed enriched consensus sequences for general transcription factors at a peak size of 200-500 bp (Figure 5-35C). In a distance of 100 bp to the Sp1-bound motifs all the other four motifs for the general transcription factors (NFY, GABP, YY1 and NRF1) as well as the unknown motif were significantly enriched. At NRF1-bound peaks, motifs for Sp1, NFY and GABP showed an enrichment with high significance in a radius of 100 bp around the bound motif. At YY1-bound peaks also with a peak size of 200 bp, consensus sites for YY1 and GABP and the unknown motif were enriched. Within a distance of ± 250 bp around the Sp1-bound motif, in addition to the other motifs, the consensus site for CREBP1 was enriched with high significance (P value: 2.7×10^{-101}) and the YY1-bound peaks were additionally co-enriched with motifs for NRF1 and vJUN within the greater distance.

Results

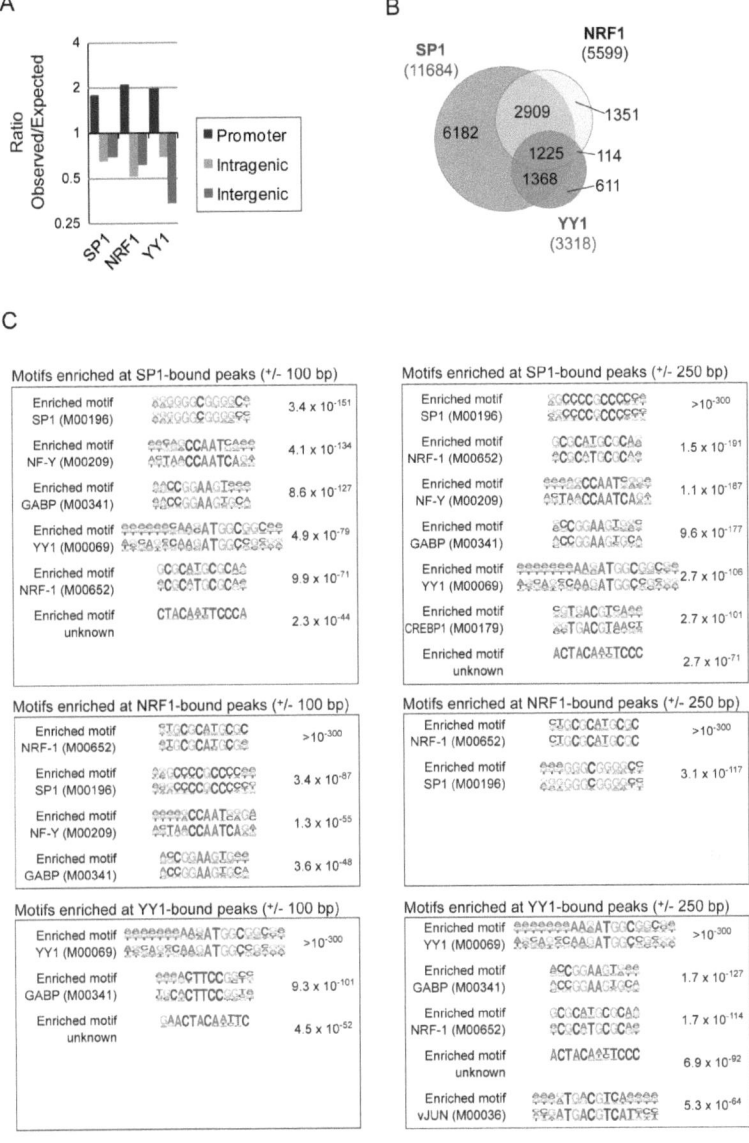

Figure 5-35 Basic analysis of ChIP-on-chip experiments for Sp1, NRF1 and YY1
(A) The distribution of binding events was analyzed dependent on their genomic location. Enrichments or depletions at the three position classes were highly significant (hypergeometric test: P<0.001). (B) The Venn diagram illustrates the overlap of bound regions between the three studied transcription factors. (Maximum distance between two peaks: 250 bp). (C) *De novo* motif analysis using the bound regions defined by ChIP-on-chip experiments. Shown are enriched motifs and corresponding TRANSFAC motifs for each transcription factor analyzed at a peak size of 200 or 500 bp.

Results

Plotting the enrichment of one of the six consensus sites for known transcription factors against the distance to a bound motif (NRF1, Sp1, YY1) reflects that some motifs showed preferences in terms of orientation or distance to each other as demonstrated in Figure 5-36. In general, motif distances show periodical preferences in most cases which comes along with sterical features caused by the helical structure of DNA. For example, YY1-bound motifs preferentially associate with NRF1, Sp1, GABP or NFY site in 5'-direction, NFY motifs are enriched at -20 bp upstream of bound Sp1 sites, and the unknown motif is preferentially located 30 bp upstream or downstream of bound NRF1 sites (Figure 5-36).

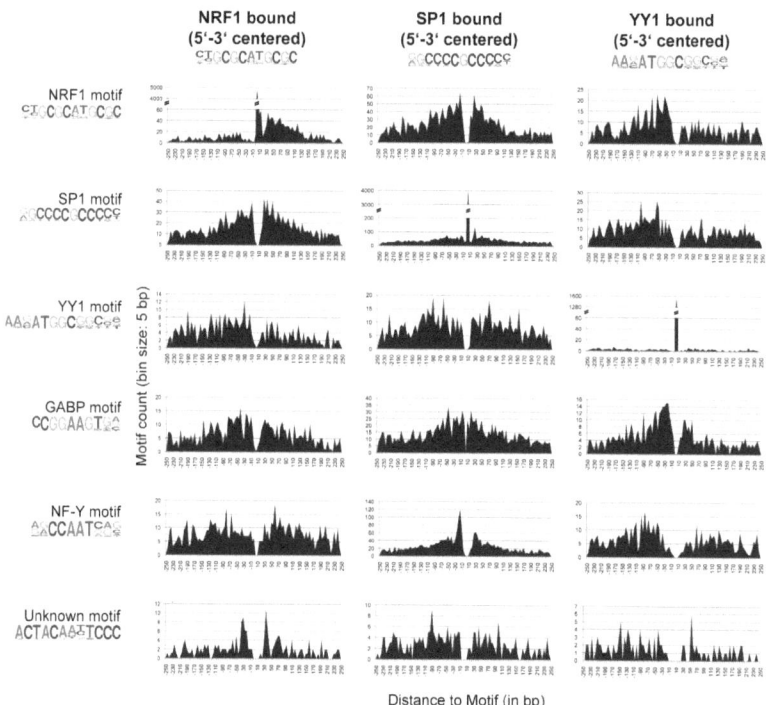

Figure 5-36 Distribution of transcription factor motifs relative to the three motifs for NRF1, Sp1 and YY1 at bound sites
YY1-bound motifs preferentially associate with NRF1, Sp1, GABP or NFY site in 5'-direction, NFY motifs are enriched at -20 bp upstream of bound Sp1 sites, and the unknown motif is preferentially located 30 bp upstream or downstream of bound NRF1 sites.

Results

The next question to be addressed was how the binding of specific factors to their consensus motif influences transcription of the respective gene. Comparing the expression data of CD34+, CD14+ and U937 cells with the ChIP-on-chip data revealed that genes associated with transcription factor-bound CpG islands generally showed significantly higher mRNA levels in CD34+ cells, CD14+ cells or the leukemia cell line as compared to all genes. The box plots showing the distribution of mRNA expression ratios are illustrated in Figure 5-37A. Moreover, the expression data were analyzed according to the number of bound transcription factors. Figure 5-37B demonstrates that binding of more factors generally increased overall expression levels of associated genes.

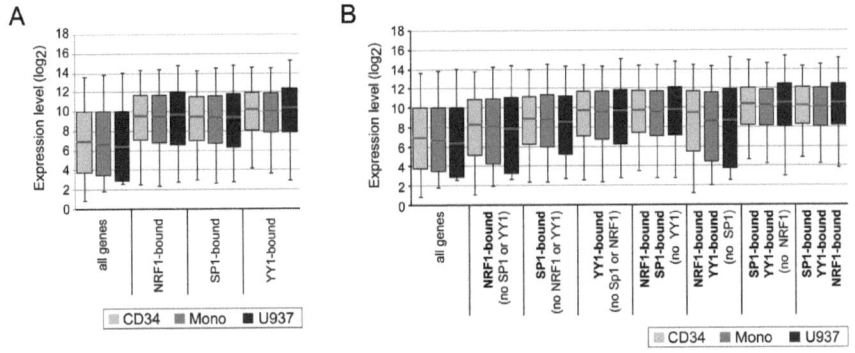

Figure 5-37 Expression status dependent on the binding of general transcription factors
(A) The box plots show the distribution of mRNA expression ratios (CD34+ progenitor cells, CD14+ normal blood monocytes, U937 cells) conditional on the binding status at individual, gene-associated peaks. The red lines denote medians, boxes the interquartile ranges, and whiskers the 5th and 95th percentiles. Pair wise comparisons of total mRNA expression ratios (all genes) and transcription factor-bound regions are significant ($P<0.001$, Mann–Whitney U test, two-sided). (B) The box plots show the distribution of mRNA expression ratios (CD34+ progenitor cells, CD14+ normal blood monocytes, U937 cells) conditional on the binding status (binding of one, two or three factors) at individual, gene-associated peaks. The red lines denote medians, boxes the interquartile ranges, and whiskers the 5th and 95th percentiles. Pair wise comparisons of total mRNA expression ratios (all genes) and transcription factor-bound regions are significant ($P<0.001$, Mann–Whitney U test, two-sided).

To directly compare transcription factor binding patterns in normal cells with aberrant methylation profiles of leukemia cell lines, the signal intensity ratios of ChIP enrichment for each transcription factor was plotted against the MCIp enrichment of the leukemia cell lines (THP-1 and U937) versus normal human blood monocytes. Figure 5-38 demonstrates that both events were mutually exclusive for all three transcription factors in U937 as well as THP-1 cells. This demonstrates that transcription factor binding protects from *de novo* methylation in leukemia cells.

Results

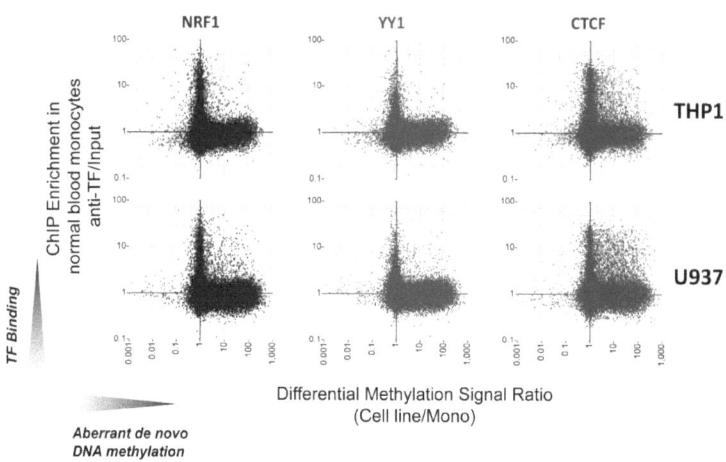

Figure 5-38 Correlation between transcription factor binding in normal cells and aberrrant *de novo* methylation in leukemia cells
The three transcription factors Sp1, NRF1 and YY1 were analyzed using ChIP-on-chip on human 244K CpG island arrays. In the diagrams the signal intensity ratios of ChIP enrichment of each transcription factor are plotted against the MCIp enrichment of the leukemia cell line (THP-1) versus normal human blood monocytes.

We also observed that transcription factor binding was not detected at every motif. Based on ChIP-on-chip data, the motifs for Sp1, NRF1 and YY1 could be subdivided into those that are not bound and those that are actually bound by the corresponding factor in CpG islands. About 35% of the Sp1 motifs, 25% of the NRF1 motifs and 16% of the YY1 motifs are not bound by the respective transcription factor. Using the *de novo* motif discovery algorithm bound and non-bound motifs were compared revealing that motifs - only if bound by transcription factors – were highly significantly co-enriched for consensus motifs of the other "protective" motifs within the distance of ± 250 bp around each motif (Figure 5-39A). Sp1-bound motifs compared to the unbound Sp1 motifs were co-enriched with motifs for NFY, NRF-1, GABP, CRE-BP1 and YY1. NRF1-bound motifs compared to the unbound NRF1-motif were co-enriched with motifs for GABP, vJUN, Sp1 and the unknown factor. The YY1-bound motif versus the unbound YY1-motif showed co-enrichment for motifs of an E-Box, Sp1, GABP and an unknown factor. P values ranged from 10^{-93} to 10^{-300}. Furthermore, ratios of observed versus expected motif occurrences were calculated for sequence motifs that are either bound by the corresponding factor or not, dependent on the number of additional consensus sequences in their close vicinity. As shown in Figure 5-39B a sequence motif was more likely bound, if it contained at least one or better two

Results

other motifs in close proximity (± 250 bp). Enrichment in the bound fraction and depletion in the unbound fraction were highly significant (in most cases hypergeometric test: P<0.001). Moreover, genes associated with transcription factor-bound motifs showed significantly higher mRNA levels as compared to genes that were associated with non-bound motifs. The mRNA expression levels in CD34+ progenitor cells, CD14+ normal blood monocytes and U937 cells conditional on the binding status of the associated motif (NRF1, Sp1, YY1) are demonstrated in Figure 5-39C.

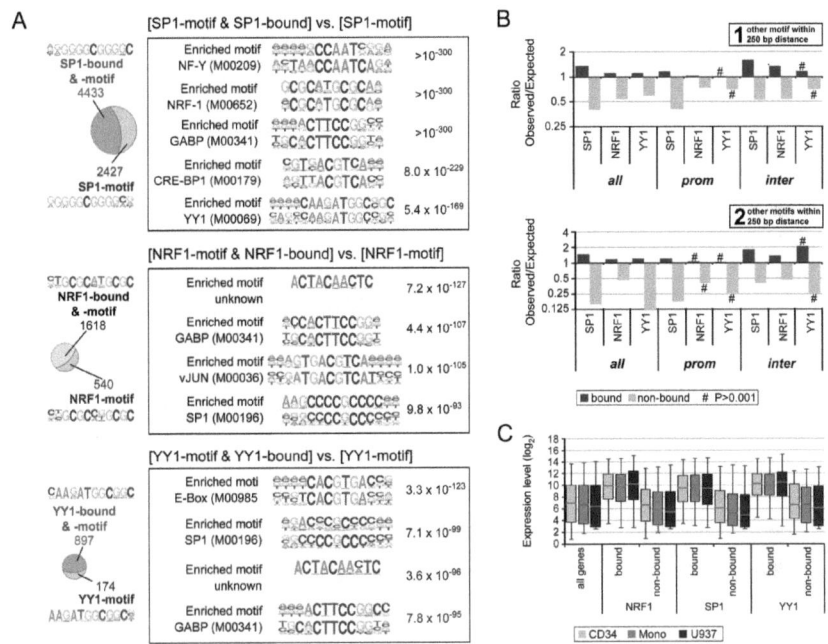

Figure 5-39 Properties of consensus sequences that are bound or not bound by the corresponding transcription factor
(A) Based on ChIP-on-chip data, the motifs for Sp1, NRF1 and YY1 could be subdivided into those that are not bound and those that are actually bound by the corresponding factor in CpG islands. *De novo* motif searches of bound motifs against non-bound motifs revealed a highly significant association of bound motifs with consensus sites for other general factors within the range of ± 250 bp around each motif. (B) Ratios of observed versus expected motif occurrences are shown for sequence motifs that are either bound by the corresponding factor (blue bars) or not bound (green bars) and had at least one (top panel) or two other consensus sites (bottom panel) within a 250 bp distance. Enrichment in the bound fraction and depletion in the unbound fraction were highly significant (hypergeometric test: P<0.001) except for the cases marked with a hash. (C) The box plots show the distribution of mRNA expression ratios (CD34+ progenitor cells, CD14+ normal blood monocytes, U937 cells) conditional on the binding status of the associated motif. The red lines denote medians, boxes the interquartile ranges, and whiskers the 5th and 95th percentiles. Pair wise comparisons of mRNA expression ratios associated with bound and non-bound motifs are significant (P<0.001, Mann–Whitney U test, two-sided).

Results

The data suggest that the stable binding of these general transcription factors (as measured by ChIP) to their consensus motif depends on the presence of neighboring motifs that are cooperatively bound by other general transcription factors. Thus, the combinatorial presence of two or more of the identified consensus sequences may serve to stabilize transcription factor binding and to confer the resistance of certain CpG islands (preferably those acting as promoters) to aberrant methylation.

5.4.6 Properties of CpG island-associated genes in conjunction with CpG island methylation status and transcription factor binding

We finally asked the question whether DNA methylation status or transcription factor binding events at CpG islands are associated with attributes or distinct properties of the corresponding genes or their products. In order to assess their biological interpretation the annotation tool in the HOMER software was used to determine enrichment or depletion of about 37,000 attributes. Since CpG islands were intrinsically enriched for many attributes as compared to the whole genome, enrichment and depletion as well as corresponding P values were calculated for individual gene groups against the total CpG island associated gene group. Hierarchical clustering of $\log_{10}$-transformed P values was performed using Genespring Software 10.0 (Agilent) using all attributes with enrichment or depletion P values $<10^{-10}$ in at least one gene list. Thirteen databases (see section 3.12) were analyzed for enrichment of specific terms or properties including gene ontology terms, pathway association, protein domains or interactions, chromosomal localization and predicted miRNA targets in regions that were associated with a particular DNA methylation status or bound by any of the three transcription factors Sp1, NRF1 or YY1 (Figure 5-40).

Hierarchical clustering of P values clearly separated the three classes of CpG islands into functional groups (Figure 5-40). DNA methylation-free and transcription factor-bound regions included properties and terms that were associated with basic cellular functions required for cell survival and proliferation. In line with earlier observations (Bracken et al., 2006a) (Figure 5-27A), CpG island regions that are commonly targeted by aberrant DNA methylation in both myeloid cell lines exhibited highly significant associations with gene ontology terms related to developmental processes, transcription factor or receptor functions, as well as homeobox proteins, that are often targeted by Polycomb group repressors. Those regions are not bound by the three general transcription factors (NRF1, Sp1, YY1). Interestingly, these associations were also found in regions that contained unbound consensus motifs for at least one of the three above general transcription factors, and to a lesser extend in regions that were methylated also in normal somatic cells (human blood monocytes). (The complete list of gene attributes and properties and the correspondent P values is given within the supplementary material of the corresponding publication.) If a region is only bound by one transcription factor alone, the protection from *de novo* methylation is very low and no significant enrichment or depletion of the CpG island regions bound by a single transcription factor within the distinct CpG island classes (the CpG island regions which remain unmethylated in normal as well as in cancer cells and those CpG island regions that become methylated during tumorigenesis) could be observed.

Results

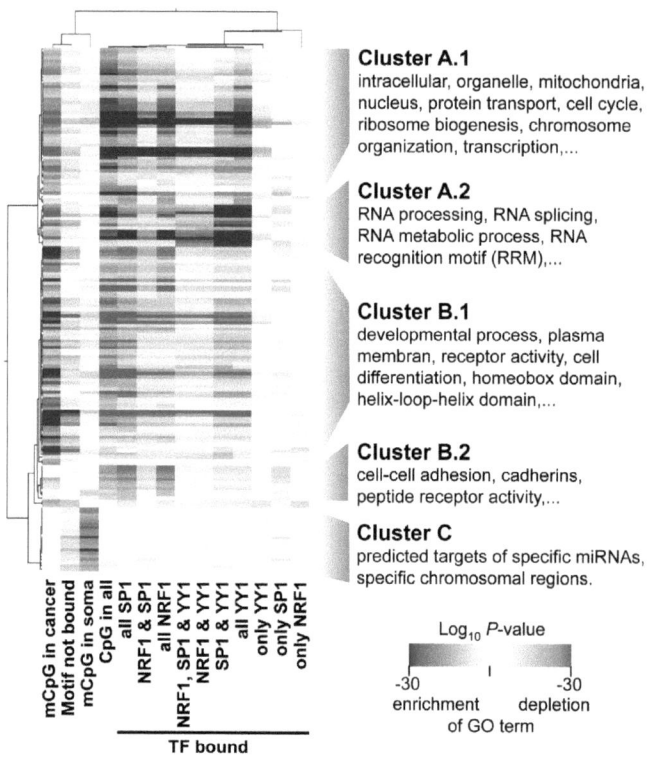

Figure 5-40 Hierarchical clustering of significance values for gene ontology enrichment
Enrichment or depletion was calculated for gene attributes and properties including gene ontology terms, pathway association, protein domains or interactions, chromosomal localization and predicted miRNA targets (a complete list of databases is given in the corresponding publication) in regions that were associated with a DNA methylation status (mCpG, methylated; CpG unmethylated), motif presence without DNA binding of the respective factor or transcription factor binding (any of the three transcription factors Sp1, NRF1 or YY1, in total (all), alone (only), or in combination). P values for enrichment or depletion of each attribute was calculated using the hypergeometric test (the complete list of P values is given within the supplementary material of the corresponding publication) and attributes with $P < 10^{-10}$ were used to perform hierarchical clustering (Pearson centered, average linkage). Data is presented as a heatmap where red coloring indicates the significant depletion and blue coloring the significant enrichment of an attribute. Main clusters of attributes are indicated and top terms are given for each group.

6 Discussion & perspectives

DNA methylation and modifications of histone tails are key cooperating mechanisms involved in maintaining epigenetic memory in mammalian cells. Along with genetic alterations, epigenetic abnormalities play an important role in gene deregulation in cancer (Jones and Baylin, 2007). Hundreds of genes show aberrant hypermethylation in specific tumor types. Initially, this was shown to be part of a silencing mechanism for tumor suppressor-like genes (Jones and Baylin, 2002), but subsequent experiments demonstrated that a large variety of different gene types (Costello et al., 2000; Yu et al., 2005) can be affected by this phenomenon. This aberrant methylation is set up early in tumor development. In order to identify potential disease markers, three cell lines as well as 25 AML and 10 colorectal carcinoma samples were screened for DNA methylation on a genome-wide level using the newly developed methyl-CpG immunoprecipitation (MCIp) approach. To get insights into the molecular basis for aberrant methylation profiles the MCIp data was analyzed using a powerful *de novo* motif search algorithm.

6.1 MCIp in comparison with existing methods

We developed a novel application allowing for the rapid and sensitive screening of DNA methylation. The central technique, called MCIp (methyl-CpG immunoprecipitation), is based on the binding of methylated DNA fragments to the bivalent, antibody-like fusion protein MBD-Fc (a methyl binding domain fused to an Fc-tail) in an immunoprecipitation-like approach. Enriched methylated DNA fragments can be efficiently detected both, on single gene level and throughout the genome. The power of this novel technique was demonstrated by the identification and subsequent validation of a large number of genes that are affected by aberrant hypermethylation in myeloid leukemias.

At present, several techniques are applied for the detection of CpG methylation (Dahl and Guldberg, 2003). Commonly used assays rely on two basic principles to distinguish methylated and unmethylated DNA: digestion with methylation-sensitive restriction enzymes or bisulfite treatment of DNA (Ammerpohl et al., 2009; Dahl and Guldberg, 2003; Frommer et al., 1992b). Approaches based on methylation-sensitive restriction enzymes enrich fragments dependent on the digestion of methylated (Irizarry et al., 2008) or unmethylated DNA (Hatada et al., 2006) followed by size fractionation. A major disadvantage of these methods is that the enzyme pairs only recognize specific sequence motifs, thus the selection of restriction enzymes automatically limits the number of detectable sequences – a global analysis of CpG methylation can therefore not be achieved (Ammerpohl et al., 2009; Dahl and Guldberg, 2003). In addition, when differences in global methylation patterns are reported between samples, it is impossible to decipher what proportion of these differences are located in promoter CpG islands (CGIs) rather than within intronic or repetitive elements. Treatment of genomic DNA with sodium bisulfite ($NaHSO_3$) overcomes this limitation and allows the analysis of virtually any CpG position within the genome. If genomic DNA is treated with sodium bisulfite, unmethylated cytosines are deaminated into uracil and transformed into thymidine residues during PCR, whereas methylated cytosines still appear as cytosines (Frommer et al., 1992b). Consequently, bisulfite treatment results in methylation dependent sequence variations of C to T after amplification. The PCR product can then be sequenced, directly or after subcloning of the amplified fragment. Direct sequencing yields information about the average methylation of a CpG site in a sample, while sequencing of cloned DNA allows the analysis of individual CpG sites on independent half strand DNA molecules (Ammerpohl et al., 2009; Dahl and Guldberg, 2003). The major disadvantage of cloning and sequencing is that a high number of clones have to be sequenced to gain reliable results. Furthermore, artifacts relating to PCR infidelity, incomplete bisulfite conversion, or erroneous bisulfite conversion of 5'-methylcytosine to thymine can significantly influence the results of this method (Ammerpohl et al., 2009; Dahl and Guldberg, 2003). Until recently, it was thought that bisulfite-treated DNA cannot be analyzed on a genome-wide level. Technical advances, in particular the next generation sequencing approaches, now enable high-throughput analysis of bisulfite sequences and the determination of CpG methylation virtually across the whole genome. However, this approach is extremely resource and labour intensive and certainly not suited for the analysis of large sample numbers.

Previous attempts to identify new genes that are differentially methylated in human disease have primarily taken candidate gene approaches relying on the use of techniques for gene-specific methylation analysis as described above. Within recent years, however, new high-throughput methods have made it possible to simultaneously analyze the methylation status of thousands of CGIs. However, most of those techniques like Restriction Landmark Genomic Scanning (RLGS) (Costello et al., 2002) or Methylated CpG Island Amplification (MCA) (Dahl and Guldberg, 2003; Smith et al., 2003) depend on

Discussion & perspectives

methylation-sensitive restriction enzymes and suffer from the same limitations as described above.

In contrast to the methods described above, which rely on a chemical reaction leading to a modification of the DNA molecules, the bases of DNA will stay unmodified when using methyl binding proteins to precipitate and thereby enrich methylated DNA (Ammerpohl et al., 2009). The utility of naturally occurring methyl-CpG binding (MBD) proteins to separate methylated and unmethylated DNA fragments is known for more than a decade. Already in 1994, the laboratory of A. Bird developed a method for enrichment of methylated DNA fragments by means of affinity chromatography using recombinant MeCP2 (Cross et al., 1994) (Cross et al., 1994). The technique has been used, improved and combined with further techniques by other groups (Brock et al., 2001; Shiraishi et al., 1999). A disadvantage of MeCP2-affinity chromatography is the large amount of genomic DNA required (50-100 µg) and the relatively time-consuming procedure. Also, a recent report by Klose et al. (Klose et al., 2005) demonstrated that MeCP2 requires an A/T run adjacent to the methylated CpG dinucleotide for efficient DNA binding, suggesting that MeCP2-affinity chromatography may be biased towards certain CpG motifs. In contrast, MBD2 showed no binding requirements or preferences in these and previous studies. Fraga et al. could show that recombinant MBD2 has a 50 to 100 times stronger affinity towards CpG-methylated DNA than recombinant MeCP2 (Fraga et al., 2003).

Therefore, we believed that the high methyl-CpG affinity of MBD2 (Fraga et al., 2003) combined with the bivalent, antibody-like structure of the recombinant MBD-Fc polypeptide could largely increase its binding capacity, thereby enabling the efficient retention of DNA fragments in dependence on their methylation degree. We could show that an unmethylated DNA fragment may be 200- to 500-fold depleted and that up to 80% of a highly methylated fragment were recovered in the high salt MCIp fraction demonstrating the high affinity of our recombinant polypeptide. The fractionation procedure works efficiently with DNA fragments obtained by restriction digest or ultrasonication (data not shown).

The properties of the recombinant MBD-Fc polypeptide allow for its application in small-scale assays requiring only little amounts of DNA (<300 ng) and therefore permit the profiling of DNA methylation of candidate genes from very limited cell numbers including biopsy samples or cells collected by laser-mediated microdissection. In addition, complete genome-wide methylation profiling is possible when a non-specific LM-PCR amplification

Discussion & perspectives

step and subsequent hybridization to microarrays are performed. The PCR step causing potential amplification bias may be omitted if sufficient starting material (2 µg DNA) is available.

At about the same time when we developed the MCIp approach, Weber et al. designed a related approach (MeDIP) using a 5-methylcytosine (5mC) antibody that requires a denaturing step before the immunoprecipitation of DNA fragments (Weber et al., 2005). Their analysis revealed only a small set of promoters being differentially methylated in a normal and a transformed cell line, suggesting that aberrant methylation of CpG island promoters in malignancy might be less frequent than previously hypothesized. In contrast to their observations, we detected a much higher percentage of differentially methylated genes, much more in line with previous estimates, using the same CpG island microarray platform (12K microarrays). This may reflect an inherent property of the cell lines used, however, may also point to a lesser sensitivity of the 5mC antibody approach as compared to our fractionated MCIp approach. A further advantage of the MCIp approach compared to the MeDIP technique is that MBD-Fc can separate the bulk of genomic DNA fragments into different fractions of increasing methylation density. This is due to the fact, that MBD-Fc recognizes the hydration of methylated DNA rather than 5mC itself (Ho et al., 2008). Methylated and unmethylated DNA fragments show differential elution behaviors from the MBD-Fc fusion protein when using increasing salt concentrations and can be fractionated according to their methylation degree. Thus, during the MCIp procedure, not only the highly methylated DNA can be enriched, but also intermediately methylated or unmethylated DNA is recovered without detectable sample loss. This allows for the simultaneous analysis of the whole range of DNA methylation density, including both hyper- and hypomethylated DNA fragments either within CpG island promoters or within non-CpG island promoters (Schilling et al., 2009; Schmidl et al., 2009). In contrast, the MeDIP approach specifically enriches for methylated fragments and is dependent on the CpG content of the analyzed fragments, resulting in a strong bias towards CpG-rich DNA regions (Keshet et al., 2006; Suzuki and Bird, 2008; Weber et al., 2007).

Recently, methods based on second generation sequencing such as 454 sequencing (Roche) or Solexa sequencing (Illumina) got into the focus of the epigenetic research (Ammerpohl et al., 2009; Lister and Ecker, 2009). As with bisulfite-modified DNA, MCIp-enriched material can be subjected to next generation sequencing technologies instead of hybridization to microarrays. The new sequencing technologies enable the global mapping of DNA methylation sites at single-base resolution. However, a high error rate is encountered when base-calling is performed with bisulfite-converted DNA, as after bisulfite conversion, the DNA being sequenced is effectively composed mainly of three bases. Since the resulting sequences are highly similar, this loss of complexity makes the subsequent aligning strategy much more difficult (Ammerpohl et al., 2009; Dahl and Guldberg, 2003; Lister and Ecker, 2009). Nevertheless, sequencing of bisulfite-converted DNA was feasible when using control lanes for autocalibration of the base-calling parameters to enable accurate base calling on the bisulfite-converted libraries (Lister and Ecker, 2009). Furthermore, in order to optimize the base calling performance, a multidimensional Gaussian mixtures model was developed (Cokus et al., 2008). Three techniques were recently used to generate bisulfite sequencing libraries compatible with next generation sequencing, namely MethylC-seq (Lister et al., 2008), BS-seq (Cokus et al., 2008) and Reduced representation BS sequencing (RRBS) (Meissner et al., 2008).

Comparable with microarrays, these sequencing technologies can also be restricted to distinct regions. Techniques that may be used prior to BS sequencing include not only binding of methylated DNA by proteins or an antibody (MCIp or MeDIP) but also capture of specific sequences by hybridization on a microarray or binding to beads in solution (Lister and Ecker, 2009). The disadvantage of these readout techniques is that they are still resource intensive compared with direct hybridization to microarrays. However, in contrast to large-scale sequencing approaches, microarrays produce data with only moderate resolution. To overcome resolution restrictions of the microarray platform in our studies the MCIp-microarray approach was combined with independent technologies (like MALDI-TOF MS) allowing the analysis of selected CGI at up to single CpG resolution. These validation experiments showed a high degree of consistency between both approaches.

6.2 Hypermethylated genes in leukemia cell lines and primary tumor samples

Aberrant hypermethylation in cancer cells may affect hundreds of CpG islands in a tumor-type specific manner (Issa, 2004; Kroeger et al., 2008). Therefore, DNA methylation patterns of a given tumor may offer important information for risk assessment, early detection and prognostic classification. Abnormal methylation patterns have been frequently described in acute myeloid leukemia (AML) (Issa, 2004; Toyota et al., 2001), and recent studies further support a crucial role for epigenetic changes in AML (Hackanson et al., 2008; Kroeger et al., 2008; Whitman et al., 2008; Wouters et al., 2007).

Discussion & perspectives

Epigenetic silencing by DNA methylation of cyclin-dependent kinase inhibitors (Herman et al., 1996; Kikuchi et al., 2002; Shen et al., 2003), DNA repair genes (Scardocci et al., 2006), apoptosis mediators (Furukawa et al., 2005; Murai et al., 2005), nuclear receptors (Liu et al., 2004; Rethmeier et al., 2006), transcription factors (Agrawal et al., 2007a), cell adhesion molecules (Roman-Gomez et al., 2003), and many other genes have already been reported (Boumber et al., 2007; Kroeger et al., 2008; Toyota et al., 2001; Youssef et al., 2004). However, most epigenetic studies in hematological neoplasms focused on the analysis of few candidate tumor suppressor genes because of the lack of suitable technologies to quantitatively evaluate DNA methylation on a genome-wide level as well as in large sample sets. Until recently, this has prevented extensive exploration of the role of DNA methylation in leukemia and its impact in diagnosis and outcome prediction.

For us, the establishment of the MCIp approach opened up new avenues towards unbiased genome-wide screening of methylated CpG islands. Using 12K CGI microarrays global methylation profiles of three leukemia cell lines were generated and a large number of gene fragments (more than 100) that are likely to be methylated in neoplastic cells could be identified. Interestingly, most genes that were detected as hypermethylated in leukemia cell lines showed extremely low or undetectable mRNA expression levels in corresponding microarray experiments. A comparison with published expression profiles for human bone marrow, CD33 positive bone marrow cells, as well as mature myeloid cells from healthy donors (http://symatlas.gnf.org/SymAtlas/; data not shown) indicates that a large proportion of these genes may not be significantly transcribed in myeloid cell types. This is in keeping with a recent study showing that a series of studied genes had low or undetectable expression levels in blood or bone marrow cells (Kroeger et al., 2008). A hypothetical (so far unknown) targeting mechanism may therefore induce CpG methylation of genes independent of their transcriptional status during cellular differentiation. Although such genes may not have a significant suppressor role in tumor development and/or progression, they may still serve as valuable biomarkers, provided that the targeting mechanism is specific for the disease.

Acute leukemia is characterized by a block of differentiation of early progenitors, which leads to the accumulation of immature cells in bone marrow and blood. The frequent mutation or downregulation of a relatively small number of transcription factors in AML patients suggests that the inactivation of transcriptional regulators may be critically involved in the malignant transformation process. Our methylation profiling of leukemia cell lines preferentially identified genes that are involved in transcriptional regulation. Half of the listed genes with an assigned molecular function (46/89) are involved in DNA-binding and transcriptional regulation, which indicates a significant over-representation. Aberrant hypermethylation of these transcription factor genes may lead to their epigenetic downregulation and likely contributes to the observed differentiation arrest in leukemia cells. This observation is in line with a previous study from Rush et al. (Rush and Plass, 2002) that investigated the methylation status of a large set of CpG islands in AML patients using RLGS and also found that a large proportion of the known methylated promoters (4/11) corresponded to genes involved in transcriptional regulation.

The list of hypermethylation targets contains several transcription factor genes, including *MAFB*, *JUN* and *KLF11* which are highly expressed in normal myeloid cells. A good tumor suppressor candidate e.g. is represented by the bZip transcription factor MAFB, which is expressed specifically in the myeloid lineage of the hematopoietic system. Its expression is upregulated successively during myeloid differentiation from multipotent progenitors to macrophages suggesting an essential role of MAFB in early myeloid and monocytic differentiation (Kelly et al., 2000b; Kelly et al., 2000a).

A major aim of this thesis was to develop methodology to enable screening of larger patient cohorts. Methylation profiling may help to clarify the pathophysiology of hypermethylation and provide new information on whether aberrant methylation of CpG islands in malignancies is random or specific and therefore help to identify new epigenetic marker genes. Experiments performed with the human CGI 12K microarrays, as described above, highlighted several technical limitations of this platform, including the presence of repetitive fragments leading to unwanted cross-hybridization events (non-specific binding which possibly gave rise to misleading results), and a relatively small number of representative genes. For global analysis of patient samples, another array platform provided by Agilent Technologies seemed to be better suited for this purpose. This array contains 244,000 probes (50-60 mer oligonucleotides) and covers about 23,000 CpG islands within coding and non-coding regions of the human genome (Agilent 244K CpG island microarrays). After improving and refining the MCIp method and its adaptation to the 244K microarray platform, the MCIp-on-chip approach was much more sensitive and provided much more information.

Using this newly adapted MCIp-on-chip approach, global methylation analysis of the cell lines was repeated. All CpG islands validated as hypermethylated in the first study (with 12K CpG island arrays) were again detected as hypermethylated in these experiments (performed with 244K Agilent arrays) provided they were included in the array design. In

Discussion & perspectives

total, approximately 11,300 or 8,700 (out of 23,000) independent regions were significantly enriched or depleted (>2.5-fold different) in U937 and THP-1, respectively. Validation using mass spectrometry analysis (1,150 amplicons covering about 140 genes) showed high consistency for both approaches.

To provide a general overview of global DNA methylation changes not only in tumor cell lines but also in primary tumor samples, we have characterized the DNA methylation profile of 25 AML patients (of primary normal karyotype) as well as 10 colorectal carcinoma patients. Major findings of this study were: (i) more than 6,000 hypermethylated CGI regions common in at least three AML patients could be identified, (ii) the analyzed AML samples showed highly variable DNA methylation for the analyzed CGI regions, (iii) tumor cell lines showed a much higher degree of methylation than primary tumors, (iv) many genes that were hypermethylated in AML samples represent PcG targets, and (v) colon DNA derived from 60-year-old healthy donors showed age-dependent methylation.

The finding that tumor cell lines showed a notably higher degree of methylation than primary tumors is probably due to the fact that cell lines often acquire additional alterations both on genetic and epigenetic levels during prolonged *in vitro* culture. It has been reported that a large proportion of genes are hypermethylated across multiple cancer cell lines, suggesting that these differences are due to intrinsic properties in generating cell lines (Smiraglia et al., 2001). The potential role of culture effects has been further highlighted by a recent study demonstrating that DNA methylation profiles of human embryonic stem cells vary over time in culture, with different genes affected in different cell lines (Allegrucci and Young, 2007).

A large number of genes hypermethylated in AML represent PcG targets which is in line with earlier studies (Grubach et al., 2008). The T-Box (TBX) 4 protein, for instance, is a classical PcG target that is often *de novo* methylated in leukemia and colorectal carcinoma but also in normal aged colon (Jin et al., 2009; Yasunaga et al., 2004). The encoded transcription factor is involved in the regulation of developmental processes and also showed a high degree of methylation in our studies: almost all AML patients were highly methylated within the promoter CpG island of this gene.

Remarkably, the methylation pattern of the normal colon tissue DNA derived from older healthy donors showed that many CpG islands methylated in AML become also methylated in healthy colon epithelium during aging. This is in line with published studies confirming that molecular changes accumulate over time with a contribution of environmental influences resulting in methylation changes as shown for the promoter regions of *MLH1* (Kurkjian et al., 2008), *ER, IGF2, N33* and *MyoD* (Ahuja and Issa, 2000) and eventual progression to colorectal cancer. Consistent with our study, Ahuja et al. reported that age-related methylation involves at least 50% of the genes which are hypermethylated in colon cancer (Ahuja and Issa, 2000). The association between aging and increased predisposition to develop cancer has long been noted (Kurkjian et al., 2008), however, there is no experimental or mechanistic evidence of a direct relationship (Fraga and Esteller, 2007). It has been demonstrated that normal aging cells and tissues show global hypomethylation (Calvanese et al., 2009), but there is also evidence for regional age-related increases in methylation of specific gene promoters (Calvanese et al., 2009) such as *RUNX3, TIG1, E-cadherin, c-fos* and *collagen alpha 1* (Fraga and Esteller, 2007). Once a critical methylation density is reached, those promoters have the potential to permanently silence gene expression (Issa, 2003).

The genes affected by hypermethylation during aging detected in this work were mainly those genes involved in developmental processes like homeobox genes or Polycomb targets. In contrast to colon, monocytes did not show age-dependent differences in methylation patterns at a large set (400) of typical *de novo* methylation targets. One explanation could be, that colon crypt stem cells may be characterized by an exceptionally high proliferation rate, resulting in a higher tendency to *de novo* DNA methylate certain CGIs.

6.3 Towards relevant disease markers for AML

Despite considerable progress during recent years, AML still remains a highly fatal disease. Many patients who already achieved complete remission relapse and die of this heterogeneous disease. The main outcome predictors of AML include age, white blood cell count and a history of a preceding malignancy. However, to complete treatment stratification, in particular for AML with normal karyotype, molecular markers are necessary. Notwithstanding the advances in molecular genetics, the current classification system does not completely reflect the heterogeneity of AML (Bullinger et al., 2009). In order to improve the molecular AML classification global analysis approaches have been applied. Expression studies already achieved considerable results by identifying novel AML subgroups and prognostic gene expression signatures (Bullinger et al., 2004; Valk et al., 2004; Verhaak et al., 2009). However, expression analyses will not be sufficient for classification and therapeutic decision making of AML. Microarray expression analyses

Discussion & perspectives

measure the abundance of mRNA, a molecule that is highly susceptible to degradation. Therefore, the standardization of microarray experiments is still challenging. In contrast, changes in DNA methylation represent a stable DNA modification which is conserved throughout sample preparation and therefore less prone to sample preparation-related changes. Thus, a DNA-based prognostic marker might provide a significant advantage to RNA-based methods (Bullinger et al., 2009). Several studies describing large-scale DNA methylation analysis to identify clinically relevant marker genes have been published recently. One publication by Martin-Subero et al. compared DNA methylation profiles of a wide range of different hematological neoplasies. Using bead arrays, they identified hypermethylation targets specific for the respective hematological tumor type as well as targets that were methylated in all hematological tumor types. But the study focused on candidate genes, selected from 807 genes, previously reported to be differentially methylated (Martin-Subero et al., 2009). Another study defined a methylation-based outcome predictor for patient survival supporting the hypothesis for possible correlations. They reported that the most predictive region comprises the promoter sequence for *KIAA1447 (BAHCC1)* (Bullinger et al., 2009). However, this DNA methylation study is again based on the analysis of specific candidate genes, but nevertheless suggests that the integration of DNA methylation data into a clinically relevant prediction model might be possible. Furthermore, the methylation of tumor suppressor genes seems to be implicated in the relapse risk of AML (Agrawal et al., 2007b; Kroeger et al., 2008). Using the HELP (*Hpa* II tiny fragment enrichment by ligation-mediated PCR) assay, Figueroa et al. performed genome-wide CGI promoter methylation studies with a set of 344 newly diagnosed primary AML samples. The large-scale epigenetic analyses revealed unique AML subgroups and methylation patterns that are associated with clinical outcome (Figueroa et al., 2010).

All these published studies point to a role of DNA methylation as a molecular biomarker. However, most of the underlying experiments either depended on restriction enzymes and therefore on specific sequence motifs or they just focused on a panel of candidate genes. Using an unbiased genome-wide approach to detect global DNA methylation combined with validation by MALDI-TOF MS, the aim of our studies was to identify the most predictive epigenetic markers in AML. Furthermore the CpG islands in our studies were not restricted to promoter regions, but also covered intragenic and non-coding intergenic regions. We decided to investigate the methylation of CGIs as the vast majority of CpG islands are usually completely unmethylated in normal tissues in both active and inactive genes (with the exception of imprinted loci and the inactive X chromosome of females) and

Discussion & perspectives

therefore do not relate to tissue-specific gene expression (Estecio and Issa, 2009; Esteller, 2002). Consequently, hypermethylation of normally unmethylated CGIs should be due to a tumor-specific event.

Our comprehensive methylation profiling led to the identification of more than 6,000 hypermethylated CGI regions common in at least three AML patients. In concordance with the heterogeneous expression patterns of AML samples (Valk et al., 2004), we detected very heterogeneous and highly variable methylation patterns throughout the analyzed AML samples. These results indicate that multiple mechanisms may operate to generate the observed epigenetic aberrations.

However, despite the overall variable patterns, a large number of genes were affected by methylation in almost all AML patients. These genes are mainly involved in transcriptional regulation and support earlier reports that point to a role of HOX and Polycomb as target genes in leukemia (Bullinger et al., 2009; Grubach et al., 2008). Besides transcriptional regulation, hypermethylation targets in our studies were also involved in cell-cell adhesion, cadherins and peptide receptor activity or age-dependent methylation as described above. Some of the detected hypermethylated genes are already known as potential candidate genes of tumors (e.g. *CDKN2B*, *CDKN2A*, *NPM2* (Kroeger et al., 2008), *SLIT2*) while most of them have not yet been described as commonly methylated genes (*SMUG1*, *ZIC1*, *MAP3K13*, *FGF12*). AML is one of the few neoplasms that show methylation of *CDKN2B* (also known as *p15/INK4B*) (Herman et al., 1997), a gene that plays an important role in TGF-β (transforming growth factor β)–induced growth arrest. In our study, the frequency of *CDKN2B* promoter methylation was relatively lower than previously reported (Herman et al., 1997), but is in line with studies performed by Toyota et al. (Toyota et al., 2001).

In order to define biomarkers specific for acute myeloid leukemia, 400 target regions (out of 6,000 regions affected by hypermethylation in AML) that are important for transcription or gene regulation or show age-dependent methylation, were chosen for screening a larger patient collection (200 AML patients) using the MassARRAY EpiTYPER approach. Both approaches (MCIp-on-chip and MALDI-TOF MS) were highly consistent and reliable results can be achieved using a combination of both techniques. The computational analysis of this data is not yet finalised. But the final objective will be to discover potential marker genes as well as correlations between methylation data and clinical parameters. Finally, such biomarkers then offer new possibilities for targeted treatment of patients and outcome prediction.

6.4 Establishing DNA methylation patterns through *cis*-acting sequences and combinatorial transcription factor binding

One of the main questions concerning CpG islands (CGIs) is why these sequences are protected from the wave of *de novo* methylation at the time of implantation when almost the entire genome undergoes *de novo* methylation, or likewise, why some CGIs become *de novo* methylated in cancer while others are protected from it. It was often assumed that this may be a function of local CpG ratio or the GC content. Some experiments however, in ES cells (Brandeis et al., 1994) and transgenic mice (Siegfried et al., 1999) indicated that CpG island methylation is controlled by specific local *cis*-acting sequences (Straussman et al., 2009) which can be bound by specific factors.

The hypothesis that a transcription factor provides methylation protection dates back to the reports of two independent groups in 1994, showing that a Sp1-binding site plays a role in protecting the adenine phosphoribosyltransferase (*APRT*) gene from *de novo* methylation in humans and mice (Brandeis et al., 1994; Macleod et al., 1994). Since Sp1-deficient animals had no obvious 'methylation defects', the concept of methylation protection by transcription factors has been controversially discussed. Likewise, binding of the insulator protein CTCF has been shown to protect a linked transgene from heterochromatin-mediated extension and subsequent *de novo* DNA methylation (Feltus et al., 2006; Mutskov et al., 2002). Indeed, CTCF can act as chromatin barrier by preventing the spread of heterochromatin structures. Furthermore CTCF binding to a differentially methylated domain upstream of the *H19* gene is required to maintain the unmethylated state and proper expression of the maternal *H19* allele.

Anecdotal evidence clearly supports a role of additional specific DNA binding proteins in establishing and maintaining DNA methylation patterns. Boumber et al., for example, described a polymorphism in the *RIL* (a candidate tumor suppressor gene) promoter that creates a Sp1/Sp3 binding site and therefore protects against methylation in cancer. Thereby it serves as direct proof that genetic polymorphisms can influence an epigenetic state (Boumber et al., 2008). Another study demonstrated that glucocorticoid hormones were found to induce stable DNA demethylation within a key enhancer of the rat liver-specific tyrosine aminotransferase (*TAT*) gene (Thomassin et al., 2001). Also other studies showed that regulation of local DNA methylation status by transcription factors could indeed provide a way to modulate gene expression during development (Han et al., 2001; Kress et al., 2006; Lin and Hsieh, 2001; Macleod et al., 1994; Tagoh et al., 2004).

However, it is still unclear whether the reported observations represent isolated cases or whether methylation protection represents a general mechanism.

Earlier computational studies identified specific nucleotide sequences that correlated with CGIs which are either prone or resistant to methylation in cancer samples. Feltus et al. identified a set of 13 sequence motifs derived from methylation-prone or methylation-resistant CGIs in multiple DNMT1 overexpressing clones using MEME and MAST algorithms. These sequence features were thought to act in *cis* to play a role in the local susceptibility of CGIs to aberrant DNA methylation (Feltus et al., 2006). Using an algorithm program, called HDFINDER, Das et al. was able to identify sequence motifs using data from normal human adult brain DNA which had similar sequence dependence on the epigenetic state of some selected CGIs as demonstrated in studies from Feltus et al. (Das et al., 2006). Studies from Keshet et al. showed a statistical enrichment of several short sequence motifs in hypermethylated promoter regions from Caco-2 and PC3 cells performing mDIP combined to microarray (containing approximately 10,000 promoter elements) analyses. Hypergeometric P values of the subsequent motif finding algorithm ranged from 10^{-4} to 10^{-9} (Keshet et al., 2006). A paper from Bock et al. demonstrated that besides sequence patterns also repeat frequencies and predicted DNA structures are highly correlated with CpG island methylation (Bock et al., 2006).

On the basis of the above computational analyses, it was postulated that most *de novo* methylation in cancer takes place in an instructive manner through interactions between *cis*-acting sequences on the DNA and *trans*-acting protein complexes capable of recruiting DNA methyltransferases. An example for this mechanism has been observed in promyelocytic leukemia: the PML-RAR fusion protein can induce gene hypermethylation and gene silencing at specific target promoters (Keshet et al., 2006).

All studies described above were based on few CpG islands and none of them was able to identify defined consensus sequence motifs resembling consensus sites for known transcription factors (Straussman et al., 2009). Only one recent survey of methylation states at CpG islands in normal human tissues described the association of unmethylated CpG islands with the consensus motif for the human zinc finger transcription factor specific protein (SP) 1 and for the signal transducer and activator of transcription (STAT) 1 transcription factor (Straussman et al., 2009).

To address the question why some CGIs are resistant to CpG methylation in cancer cells while others are prone to methylation, we used the global methylation profiles generated by the newly developed and adapted MCIp-microarray (MCIp-on-chip) approach. Using a powerful *de novo* motif analysis (HOMER) it could be shown that a number of defined sequence motifs are strongly enriched in CpG islands that are generally resistant to *de novo* methylation in cancer. These sequence motifs were previously shown to represent the most conserved motifs in mammalian promoters such as NRF-1, NFY, Sp1 and GABP (Xie et al., 2005a). However, the observed correlation is also evident at intergenic, promoter-distal CpG islands that are not directly associated with transcription.

We also showed that the sole presence of a consensus motif for any of the general factors is not sufficient to confer 'protection' from *de novo* methylation. In fact, protection from *de novo* methylation requires the stable binding of these factors to their binding sites which, in turn, requires the presence of neighboring motifs that are co-bound by at least one other ubiquitous (or in some cases cell type-specific) transcription factor. The stable binding of these factors likely recruits co-factors that in turn create a protective chromatin environment, e.g. by introducing protective histone marks like H3K4 methylation. A schematic model describing the methylation protection hypothesis is shown in Figure 6-1.

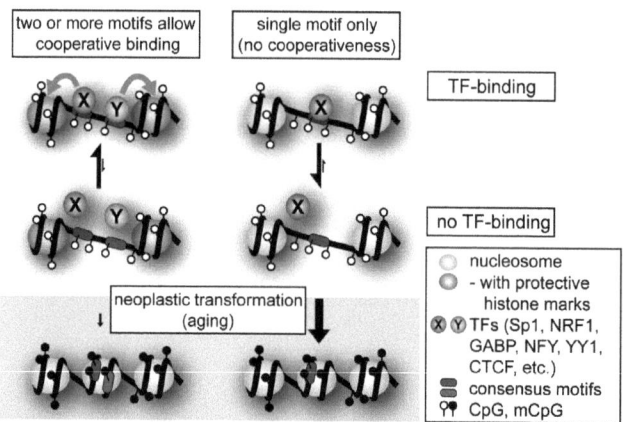

Figure 6-1 A model for DNA methylation protection by the combinatorial action of general transcription factors
If two or more consensus sites for general transcription factors are located in close proximity, these sites are likely to be bound stably by the corresponding factors. The stable binding of these factors likely recruits co-factors that in turn create a protective chromatin environment, e.g. by introducing protective histone marks like H3K4 methylation. These regions are only rarely methylated during neoplastic transformation or aging. A single, isolated motif is less likely to be bound by its corresponding factor and will have a less protective chromatin environment. These regions are more likely targeted by *de novo* methylation in cancer.

Most methylation-resistant CpG islands were bound by combinations of ubiquitous transcription factors and were also associated with attributes associated with basic cellular functions like cell survival and proliferation, whereas methylation-prone CpG islands generally associated with organismic development, differentiation and cell communication, which are frequently regulated by cell type-specific transcription factors. A schematic model which describes the different role of ubiquitous transcription factors compared to cell type-specific transcription factors with regard to protection from *de novo* methylation is shown in Figure 6-2.

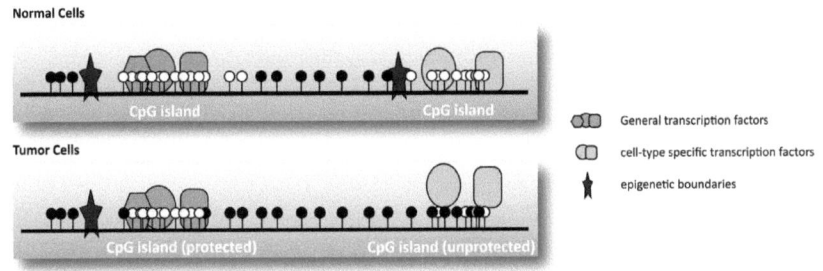

Figure 6-2 Transcription factors protect from *de novo* methylation
Protective motifs are bound by general transcription factors (marked in blue) creating a chromatin environment that excludes DNA methylation. Therefore the probability of acquiring *de novo* methylation is permanently low. In contrast, cell type-specific transcription factors (marked in yellow) may only offer temporary protection (e.g. during embryonic development) and have an increased probability of acquiring *de novo* methylation over time. The loss of epigenetic boundaries (marked as stars) (e.g. CTCF) may further increase probability of acquiring *de novo* methylation.

Interestingly, genes that are associated with CpG islands that were commonly methylated in normal and cancer cells were enriched for predicted targets of specific (mostly uncharacterized) miRNAs (Figure 5-40), however, the relevance of this observation is uncertain and requires functional validation.

We also observed that methylation-prone regions are significantly enriched for certain repeat motifs (GAGA, CACA) implying that they may also act as *cis*-acting sequences and direct *de novo* DNA methylation. GAGA resembles the consensus motif for *Drosophila* GAGA-binding factor, a trithorax group member that has been implicated in preventing heterochromatin spreading (Nakayama et al., 2007), however, a mammalian homologue has not been described so far. CA-repeats may play a role in RNA splicing and are bound by the heterogeneous nuclear ribonucleoprotein (hnRNP) L in a repeat length dependent manner (Hui et al., 2003), but there is no known link to DNA methylation or chromatin structure.

Discussion & perspectives

With the exception of the Sp1/3 motif, none of the other motifs has previously been associated with the establishment or maintenance of DNA methylation (Boumber et al., 2008; Brandeis et al., 1994; Straussman et al., 2009) but all are known to recruit epigenetic modifiers to their binding sites. NFY (also known as CAAT-binding factor), a regulator of many cell cycle control genes, actively recruits co-activators (like p300) that induce histone acetylation at NFY-bound promoters (Faniello et al., 1999). Ubiquitously expressed NRF1 and GABP (also called NRF2) are able to recruit co-activators (PCG1, p300/CBP) that create a chromatin environment favoring transcription (Izumi et al., 2003). YY1 has been shown to recruit Polycomb group proteins that control H3K27 methylation, a mark that is established on unmethylated CGI genes early in development and then maintained in differentiated cell types by the presence of an EZH2-containing Polycomb complex. In cancer cells, as opposed to normal cells, the presence of this complex brings about the recruitment of DNA methyltransferases, leading to *de novo* methylation and therefore to aberrant silencing during tumorigenesis (Schlesinger et al., 2007; Vire et al., 2006). However, a recent study by Lindroth et al. elegantly demonstrated that H3K27 methylation (recruited by YY1) and CpG DNA methylation at the murine *Rasgrf1* locus are mutually exclusive, suggesting that both epigenetic marks are interdependent and antagonistic (Lindroth et al., 2008). This is also consistent with a recent study globally mapping key histone modifications and subunits of Polycomb-repressive complexes 1 and 2 (PRC1 and PRC2) in ES cells (Ku et al., 2008). PRC2 contains EZH2, which catalyzes H3K27me3. PCR1 components, in turn, contain proteins with affinity for H3K27me3. Genome-wide analysis of PRC1 and PRC2 occupancy identified a YY1-like motif enriched in CpG islands that were not targeted by PRC2. Additional motifs identified in this study (ETS, NFY, AP-1, MYC and NRF1) (Ku et al., 2008), partially overlapped with those observed in the present study. Motifs enriched in EZH2 negative CGIs are recognized by several well-characterized classes of transcriptional activators that are highly enriched in ES cells. Some of the implicated factors have key functions in the ES cell regulatory network (e.g. NFY, Myc) while others are constitutive activators with general housekeeping functions (e.g. Ets1). In contrast, in PRC2-positive CGIs transcriptional activator motifs are depleted while repressor motifs are enriched. Thus, PCR2 appears to localize to CGIs that are transcriptional silent in ES cells because they lack activating DNA sequence motifs. These findings further corroborate the negative correlation of repressive epigenetic marks and *cis*-acting sequences conferring transcriptional activity.

Discussion & perspectives

In line with several recent observations demonstrating that the DNA methylation status correlates with histone modifications (Brunner et al., 2009; Meissner et al., 2008; Schmidl et al., 2009), the factors binding the identified sequences likely share the ability to recruit RNA polymerase II and to create an 'active' chromatin environment that may prevent or at least impede *de novo* CpG methylation at particular CpG islands (Figure 6-1). A recently published, analogous study demonstrated that the presence of RNA polymerase II, active or stalled, predicts the epigenetic fate of promoter CpG islands in cancer (Takeshima et al., 2009). Through performing chromatin immunoprecipitation combined to microarray hybridization (ChIP-on-chip) analysis of RNA polymerase II (Pol II) and histone modifications it could be shown that even among the genes with low transcription, the presence of Pol II was associated with marked resistance to DNA methylation while H3K27me3 was associated with increased susceptibility (Takeshima et al., 2009).

RNA polymerase II does not stably bind DNA on its own – its stable recruitment requires *cis*-acting factors of which Sp1 is one of the best studied so far (Lemon and Tjian, 2000). A high level of overlap between transcription factor and Pol II binding is expected and the association of Pol II with resistance to *de novo* methylation is likely a consequence of its interaction with combinations of transcription factors present at the promoter. However, our data also clearly shows that the correlation of motif occurrence or transcription factor binding and 'methylation protection' are found not only in promoter proximal but also in promoter distal sites, thus ruling out a direct link with the process of transcription or the presence of Pol II. Therefore the present thesis suggests that *cis*-acting factors may have a protective role independent of Pol II binding.

In conclusion, these data provide strong experimental and computational evidence that specific sequence motifs are associated with the DNA methylation states of CpG islands in normal and malignant cells. Most of the identified sequence motifs are identical to consensus motifs for known, general transcription factors and our data strongly suggests that the combinatorial binding of these factors plays a dominant role in regulating the DNA methylation status at a large set of CpG islands. These findings also imply that the aberrant methylation patterns in cancer cells may at least in part result from a 'loss of protection'. This would also imply a default tendency to methylate and repress DNA sequences during successive cell divisions that are not marked by activating transcription factors or histone modifications.

6.5 Perspectives

The results of the present thesis led to the identification of hundreds of hypermethylated genes of potential pathogenic relevance in cancer development. Comparing the methylation patterns of the different patients should highlight correlations between methylation of specific genes and clinical parameters such as subclasses or prognosis. The final aim of our studies is the identification of specific marker genes that in the future may provide a novel basis for improved patient outcome prediction, prognostication, diagnosis, monitoring and treatment.

However, additional studies will be necessary to find an optimal set of epigenetic biomarkers and evaluate the significance of these markers in a routine clinical setting before the clinical implementation will become accomplishable. A new method with this potential would be to screen patient samples by multiplexing (12-30 plex) of an optimal set consisting of 12-30 biomarkers using QGE after MCIp enrichment. This quantitative and multiplexed methylation analyses should be much more sensitive than the methylation-specific PCR (MSP), which was used for the detection of tumor-related DNA methylation in serum/plasma, urine and other fluids. Thus, MCIp combined with QGE might be helpful to identify patient subgroups that are likely to benefit from demethylation therapy. Moreover, in future established methylation markers might be used to detect therapeutic success of demethylating agents during the course of treatment.

DNA methylation is an early event that often precedes the appearance of a tumor. As the early stages of cancer development have the highest potential for therapeutic interventions, the inhibition or the withdrawal of these epigenetic modifications could open up new possibilities for cancer prevention in the future. Furthermore, as DNA methylation possibly also changes during the course of the disease, integrated approaches could be superior for outcome prediction. A combination of methylation and gene expression markers as well as known prognostic factors such as cytogenetics and molecular alterations could account for refining AML classification.

Our global and locus-wide analyses of DNA methylation patterns strongly suggest that the combinatorial binding of *cis*-acting transcription factors plays a major role in shaping a cells' methylome, both in health and disease. Proximal promoter regions that are often studied in the context of cancer may reflect only a small proportion of regulatory regions that are subject to alterations in cancer. In order to understand the relevance of alterations in transcription factor networks for the establishment of global DNA methylation patterns, we probably need to study not only CGIs, but basically all regions within the whole range of CpG densities, because many transcription factors (like C/EBPs, RUNX1, or PU.1) do not have preferences for CpG islands. One possibility to reduce the candidate sequences from the complete human genome to potentially regulatory relevant regions would be to define putative regulatory sites by mapping histone H3 lysine 4 mono-methylation (H3K4me1) as this histone mark is often associated with enhancers. Defining the methylation profiles of those regions could allow the identification of further cis-acting sequences and corresponding transcription factors associated with differentiation and disease states.

Another future study could include the characterization of the exact mechanism establishing and maintaining the DNA methylation patterns during leukemogenesis. Using a knockdown test sytem our hypothesis that transcription factors normally confer methylation protection could be corroborated. Since DNA methylation, in particular within CGIs, may be a consequence of the absence or inactivation of transcriptional activators, knockdown of transcription factors should then lead to methylation of the respective CGI. In a complementary approach, the epigenetic profile of stably introduced plasmids into THP-1 cells containing CGIs of varying motif composition could be studied over time. These experiments could show whether certain motifs actually do confer methylation protection to the surrounding sequences. If the expected changes are reproducible, the exact timing of DNA methylation changes and other associated epigenetic events (like the recruitment of DNMTs, the loss of activating or the deposition of repressive histone marks) could be studied sequentially.

7 Summary

Aberrant DNA methylation of CpG islands (CGIs) is a common alteration during malignant transformation that leads to the abnormal silencing of tumor suppressor genes and plays a role in disease initiation and progression. The major aim of the present thesis was the implementation of methodologies to identify epigenetic marker genes that can be used for the diagnosis as well as for the targeted treatment of tumors. Furthermore, the molecular mechanisms controlling the methylation status of CpG islands in normal and malignant cells should be analyzed. To address these issues, a novel and robust technique, called methyl-CpG immunoprecipitation (MCIp) was developed that allows for the unbiased genome-wide profiling of CpG methylation in DNA samples where quantity is limited. This approach is based on a recombinant, antibody-like protein that efficiently binds native CpG-methylated DNA and enables the fractionation of DNA fragments depending on the particular methyl-CpG content. This application facilitates the monitoring of CpG island methylation either on single gene or on genome-wide levels. Initial genome-wide methylation profiling of myeloid leukemia cell lines using 12K CpG island microarrays identified over one hundred genes with aberrantly methylated CpG islands. Interestingly, the comparison with gene expression data revealed that more than half of the identified genes were not expressed in various healthy cell types, indicating that hypermethylation in cancer may be largely independent of the transcriptional status of the affected gene. The majority of individually tested genes were also hypermethylated in primary blast cells from AML patients.

The MCIp approach was further optimized and adapted for a more suitable microarray platform (Agilent 244K CGI microarrays). The in-depth comparison of MCIp and MassARRAY for two established cell lines showed an excellent correlation over a set of 140 genes (1,150 amplicons covering approximately 13,500 CpG dinucleotides). In order to identify potential marker genes, global comparative CpG island methylation profiles for more than 25 AML samples (of mostly normal karyotype) and ten patients with colorectal carcinoma using MCIp in combination with microarray were generated. Our comprehensive analysis identified a large array of CGIs that are previously unrecognized targets of hypermethylation in AML. For the identification of potential marker genes, approximately 400 regions were selected based on the array results for screening a large set of 200 AML patients. The data are now ready to be subjected to computational analyses.

Summary

In order to get insights into the process regulating the methylation status of CpG islands, factors should be identified that are responsible for maintaining or establishing methylated states of CGIs in health and disease as well as for *de novo* methylation in cancer. *De novo* motif discovery analysis revealed two repetitive sequence motifs (GAGA, CACA) that were commonly enriched in CpG islands that were methylated in cancer. More strikingly, the global analysis demonstrated a highly significant association of unmethylated CpG islands with consensus sequences for GA binding protein (GABP), specific protein (Sp) 1 and 3, nuclear respiratory factor (NRF) 1, nuclear factor (NF) Y, yin-yang (YY) 1 and an unknown factor in all analyzed samples. Using ChIP-on-chip assays we also showed that most of the identified motifs for Sp1, NRF1 and YY1 were actually bound by the respective factors in normal cells and that these regions did not acquire *de novo* methylation in leukemia cells. In addition, the data provide global evidence that the stable binding of any of these transcription factors to their consensus motif depends on their co-occurrence with neighboring consensus motifs. Thus, the results of the present thesis suggest a major role for cooperative transcription factor binding in maintaining the unmethylated status of CpG islands in health and disease. The data also implies that the majority of *de novo* methylated CpG islands are characterized by the lack of sequence motif combinations and the absence of activating transcription factor binding.

8 Zusammenfassung

Tumorzellen zeichnen sich häufig durch ein verändertes DNA-Methylierungsmuster aus. Fehlerhafte DNA-Methylierung von CpG-Inseln (CGIs) kann zur abnormen Repression von Tumorsuppressorgenen führen und Tumorwachstum fördern. Hauptziel der vorliegenden Arbeit war es, Methoden zu etablieren, um diagnostisch oder therapeutisch verwertbare epigenetische Biomarker zu identifizieren. Desweiteren sollten die molekularen Mechanismen analysiert werden, die den Methylierungsstatus von CGIs sowohl in gesunden als auch in entarteten Zellen regulieren.

Für den unvoreingenommenen, globalen Nachweis von differentieller genomischer CpG-Methylierung wurde eine neuartige Methode, die sogenannte Methyl-CpG-Immunpräzipitation (MCIp), entwickelt und etabliert. Diese Technologie basiert auf einem rekombinanten Antikörper-ähnlichen Protein, das doppelsträngige, methylierte DNA binden kann und eine Fraktionierung der DNA-Fragmente hinsichtlich ihres Methylierungsgrades ermöglicht. Die Detektion methylierter DNA kann sowohl auf Einzelgenebene als auch genomweit durchgeführt werden. Die ersten genomweiten Methylierungsanalysen von myeloischen Leukämiezelllinien mit 12K CpG-Insel-Mikroarrays führten zur Identifizierung von über einhundert Genen, die in den Zelllinien im Vergleich zu normalen Blutmonozyten von Hypermethylierung betroffen waren. Ein Vergleich mit Expressionsdaten zeigte, dass ein Großteil der methylierten Gene weder in normalen myeloischen Zellen noch in den untersuchten Tumorzellen exprimiert war. Dies könnte darauf hindeuten, dass die tumorspezifische Hypermethylierung unabhängig vom transkriptionellen Status eines Gens ist. Die meisten der getesteten Genfragmente waren auch in primären AML-Blasten hypermethyliert.

Die MCIp-Technik wurde weiter optimiert und auf eine neue und besser geeignete Mikroarray-Plattform angepasst (Agilent 244K CpG-Insel Mikroarrays). Die Validierung der Mikroarraydaten mittels MassARRAY-Technologie (1150 Amplikons aus 140 Genen, welche 13500 CpG Dinukleotide abdeckten) zeigte eine sehr gute Korrelation beider Methoden. Zur Identifizierung von potentiellen Biomarkern wurden globale DANN-Methylierungsprofile einerseits von Blasten aus 25 AML-Patienten mit primär normalem Karyotyp, aber auch von zehn Patienten mit kolorektalem Karzinom erstellt. Unsere Analysen identifizierten eine Reihe von Genen von denen bislang nicht bekannt war, dass sie von Hypermethylierung betroffen sein können. Um relevante Markergene zu identifizieren, wurden ca. 400 Regionen anhand der Arrayergebnisse ausgewählt und in

Zusammenfassung

einem größeren Patientenkollektiv (200 AML Proben) mithilfe der MassARRAY-Technologie validiert. Die entsprechenden Daten werden aktuell noch bioinformatischen Analysen unterzogen.

Um Einblick in den Mechanismus zu gewinnen, wie die Methylierung von CpG-Inseln reguliert wird, sollten Faktoren identifiziert werden, welche einen entscheidenden Einfluss bei der Entstehung und Aufrechterhaltung von Methylierungsmustern sowohl in gesunden als auch in Tumorzellen haben. Mittels *de novo*-Motivanalysen konnte gezeigt werden, dass zwei repetitive Sequenzmotive (GAGA, CACA) häufig in CGIs angereichert waren, welche in Tumorzellen methyliert wurden. Darüber hinaus stellten wir mittels globaler Analysen eine hochsignifikante Assoziation von unmethylierten CGIs mit Konsensussequenzen für GABP (*GA binding protein*), Sp1 (*Specific protein 1*), NRF1 (*nuclear respiratory factor 1*), NFY (*nuclear factor Y*), YY1 (*ying-yang 1*) und einem unbekannten Faktor in allen untersuchten Proben fest.

Mittels ChIP-on-Chip Analysen konnte außerdem gezeigt werden, dass die meisten der identifizierten Motive für Sp1, NRF1 und YY1 tatsächlich von dem betreffenden Faktor in normalen Zellen gebunden wurden, und dass diese Regionen in Leukämiezellen nicht von einer *de novo*-Methylierung betroffen waren. Desweiteren verdeutlichten die Ergebnisse, dass die stabile Bindung eines dieser Transkriptionsfaktoren an seine Konsensussequenz vom gleichzeitigen Vorkommen benachbarter Konsensusmotive abhängig ist. Folglich führen die Ergebnisse dieser Dissertation zu der Annahme, dass die kooperative Bindung von Transkriptionsfaktoren eine entscheidende Rolle für die Aufrechterhaltung des unmethylierten Status von CGIs in gesunden wie auch in kranken Zellen spielt. Die Daten implizieren auch, dass die Mehrheit der *de novo* methylierten CGIs durch das Fehlen von Kombinationen bestimmter Sequenzmotive und der daraus resultierenden Abwesenheit aktivierender Transkriptionsfaktoren charakterisiert ist.

9 References

Ades,S. (2009). Adjuvant Chemotherapy for Colon Cancer in the Elderly: Moving From Evidence to Practice. Oncology-New York *23*, 162-167.

Agrawal,S., Hofmann,W.K., Tidow,N., Ehrich,M., van den Boom,D., Koschmieder,S., Berdel,W.E., Serve,H., and Muller-Tidow,C. (2007a). The C/EBP delta tumor suppressor is silenced by hypermethylation in acute myeloid leukemia. Blood *109*, 3895-3905.

Agrawal,S., Unterberg,M., Koschmieder,S., zur Stadt,U., Brunnberg,U., Verbeek,W., Buchner,T., Berdel,W.E., Serve,H., and Muller-Tidow,C. (2007b). DNA methylation of tumor suppressor genes in clinical remission predicts the relapse risk in acute myeloid leukemia. Cancer Research *67*, 1370-1377.

Ahuja,N. and Issa,J.P.J. (2000). Aging, methylation and cancer. Histology and Histopathology *15*, 835-842.

Allegrucci,C. and Young,L.E. (2007). Differences between human embryonic stem cell lines. Human Reproduction Update *13*, 103-120.

Allis,C.D., Jenuwein,T., and Reinberg,D. (2007). Epigenetics. (New York: John Inglis).

Ammerpohl,O., Martin-Subero,J.I., Richter,J., Vater,I., and Siebert,R. (2009). Hunting for the 5th base: Techniques for analyzing DNA methylation. Biochimica et Biophysica Acta-General Subjects *1790*, 847-862.

Antequera,F. and Bird,A. (1993). Number of Cpg Islands and Genes in Human and Mouse. Proceedings of the National Academy of Sciences of the United States of America *90*, 11995-11999.

Appelbaum,F.R., Rowe,J.M., Radich,J., and Dick,J.E. (2001). Acute myeloid leukemia. Hematology. Am. Soc. Hematol. Educ. Program. 62-86.

Avner,P. and Heard,E. (2001). X-chromosome inactivation: Counting, choice and initiation. Nature Reviews Genetics *2*, 59-67.

Ballestar,E. and Wolffe,A.P. (2001). Methyl-CpG-binding proteins. Targeting specific gene repression. Eur. J. Biochem. *268*, 1-6.

Barash Y., Bejerano G., and Friedman N. (2001). A Simple Hyper-Geometric Approach for Discovering Putative Transcription Factor Binding Sites. In WABI '01: Proceedings of the First International Workshop on Algorithms in Bioinformatics, (London, UK: Springer-Verlag), pp. 278-293.

Barski,A., Cuddapah,S., Cui,K., Roh,T.Y., Schones,D.E., Wang,Z., Wei,G., Chepelev,I., and Zhao,K. (2007). High-resolution profiling of histone methylations in the human genome. Cell. *129*, 823-837.

Bartova,E., Krejci,J., Harnicarova,A., Galiova,G., and Kozubek,S. (2008). Histone modifications and nuclear architecture: A review. Journal of Histochemistry & Cytochemistry *56*, 711-721.

Bell,A.C., West,A.G., and Felsenfeld,G. (1999). The protein CTCF is required for the enhancer blocking activity of vertebrate insulators. Cell *98*, 387-396.

Bennett,J.M., Catovsky,D., Daniel,M.T., Flandrin,G., Galton,D.A.G., Gralnick,H.R., and Sultan,C. (1976). Proposals for Classification of Acute Leukemias. British Journal of Haematology *33*, 451-&.

Berger,S.L. (2007). The complex language of chromatin regulation during transcription. Nature. *447*, 407-412.

Bernstein,B.E., Meissner,A., and Lander,E.S. (2007). The mammalian epigenome. Cell. *128*, 669-681.

Bernstein,B.E., Mikkelsen,T.S., Xie,X.H., Kamal,M., Huebert,D.J., Cuff,J., Fry,B., Meissner,A., Wernig,M., Plath,K., Jaenisch,R., Wagschal,A., Feil,R., Schreiber,S.L., and Lander,E.S. (2006). A bivalent chromatin structure marks key developmental genes in embryonic stem cells. Cell *125*, 315-326.

Bernstein,E. and Allis,C.D. (2005). RNA meets chromatin. Genes Dev. *19*, 1635-1655.

Bhattacharya,S.K., Ramchandani,S., Cervoni,N., and Szyf,M. (1999). A mammalian protein with specific demethylase activity for mCpG DNA. Nature *397*, 579-583.

Bhaumik,S.R., Smith,E., and Shilatifard,A. (2007). Covalent modifications of histones during development and disease pathogenesis. Nature Structural & Molecular Biology *14*, 1008-1016.

Bird,A. (2002). DNA methylation patterns and epigenetic memory. Genes Dev. *16*, 6-21.

Bird,A.P. and Wolffe,A.P. (1999). Methylation-induced repression--belts, braces, and chromatin. Cell. *99*, 451-454.

Bock,C., Paulsen,M., Tierling,S., Mikeska,T., Lengauer,T., and Walter,J. (2006). CpG island methylation in human lymphocytes is highly correlated with DNA sequence, repeats, and predicted DNA structure. PLoS. Genet. *2*, e26.

References

Boumber,Y.A., Kondo,Y., Chen,X., Shen,L., Guo,Y., Tellez,C., Estecio,M.R.H., Ahmed,S., and Issa,J.P.J. (2008). An Sp1/Sp3 Binding Polymorphism Confers Methylation Protection. Plos Genetics 4.

Boumber,Y.A., Kondo,Y., Chen,X.Q., Shen,L.L., Gharibyan,V., Konishi,K., Estey,E., Kantarjian,H., Garcia-Manero,G., and Issa,J.P.J. (2007). RIL, a LIM gene on 5q31, is silenced by methylation in cancer and sensitizes cancer cells to apoptosis. Cancer Research 67, 1997-2005.

Bracken,A.P., Dietrich,N., Pasini,D., Hansen,K.H., and Helin,K. (2006a). Genome-wide mapping of Polycomb target genes unravels their roles in cell fate transitions. Genes Dev. 20, 1123-1136.

Bracken,A.P., Dietrich,N., Pasini,D., Hansen,K.H., and Helin,K. (2006b). Genome-wide mapping of Polycomb target genes unravels their roles in cell fate transitions. Genes & Development 20, 1123-1136.

Brandeis,M., Frank,D., Keshet,I., Siegfried,Z., Mendelsohn,M., Nemes,A., Temper,V., Razin,A., and Cedar,H. (1994). Sp1 elements protect a CpG island from de novo methylation. Nature. 371, 435-438.

Brock,G.J., Huang,T.H., Chen,C.M., and Johnson,K.J. (2001). A novel technique for the identification of CpG islands exhibiting altered methylation patterns (ICEAMP). Nucleic Acids Res. 29, E123.

Brunner,A.L., Johnson,D.S., Kim,S.W., Valouev,A., Reddy,T.E., Neff,N.F., Anton,E., Medina,C., Nguyen,L., Chiao,E., Oyolu,C.B., Schroth,G.P., Absher,D.M., Baker,J.C., and Myers,R.M. (2009). Distinct DNA methylation patterns characterize differentiated human embryonic stem cells and developing human fetal liver. Genome Res. 19, 1044-1056.

Bullinger,L., Dohner,K., Bair,E., Frohling,S., Schlenk,R.F., Tibshirani,R., Dohner,H., and Pollack,J.R. (2004). Use of gene-expression profiling to identify prognostic subclasses in adult acute myeloid leukemia. New England Journal of Medicine 350, 1605-1616.

Bullinger,L., Ehrich,M., Dohner,K., Schlenk,R.F., Dohner,H., Nelson,M.R., and van den,B.D. (2009). Quantitative DNA-methylation predicts survival in adult acute myeloid leukemia. Blood.

Calvanese,V., Lara,E., Kahn,A., and Fraga,M.F. (2009). The role of epigenetics in aging and age-related diseases. Ageing Research Reviews 8, 268-276.

Chim,C.S., Wong,A.S., and Kwong,Y.L. (2003). Epigenetic inactivation of INK4/CDK/RB cell cycle pathway in acute leukemias. Ann. Hematol. 82, 738-742.

Choi,M.C., Jong,H.S., Kim,T.Y., Song,S.H., Lee,D.S., Lee,J.W., Kim,T.Y., Kim,N.K., and Bang,Y.J. (2004). AKAP12/Gravin is inactivated by epigenetic mechanism in human gastric carcinoma and shows growth suppressor activity. Oncogene 23, 7095-7103.

Cokus,S.J., Feng,S.H., Zhang,X.Y., Chen,Z.G., Merriman,B., Haudenschild,C.D., Pradhan,S., Nelson,S.F., Pellegrini,M., and Jacobsen,S.E. (2008). Shotgun bisulphite sequencing of the Arabidopsis genome reveals DNA methylation patterning. Nature 452, 215-219.

Costa,F.F. (2008). Non-coding RNAs, epigenetics and complexity. Gene 410, 9-17.

Costello,J.F., Fruhwald,M.C., Smiraglia,D.J., Rush,L.J., Robertson,G.P., Gao,X., Wright,F.A., Feramisco,J.D., Peltomaki,P., Lang,J.C., Schuller,D.E., Yu,L., Bloomfield,C.D., Caligiuri,M.A., Yates,A., Nishikawa,R., Su,H.H., Petrelli,N.J., Zhang,X., O'Dorisio,M.S., Held,W.A., Cavenee,W.K., and Plass,C. (2000). Aberrant CpG-island methylation has non-random and tumour-type-specific patterns. Nat. Genet. 24, 132-138.

Costello,J.F. and Plass,C. (2001). Methylation matters. J. Med. Genet. 38, 285-303.

Costello,J.F., Smiraglia,D.J., and Plass,C. (2002). Restriction landmark genome scanning. Methods 27, 144-149.

Cross,S.H., Charlton,J.A., Nan,X., and Bird,A.P. (1994). Purification of CpG islands using a methylated DNA binding column. Nat. Genet. 6, 236-244.

Dahl,C. and Guldberg,P. (2003). DNA methylation analysis techniques. Biogerontology. 4, 233-250.

Daniel,J.A., Pray-Grant,M.G., and Grant,P.A. (2005). Effector proteins for methylated histones: an expanding family. Cell Cycle. 4, 919-926.

Das,R., Dimitrova,N., Xuan,Z., Rollins,R.A., Haghighi,F., Edwards,J.R., Ju,J., Bestor,T.H., and Zhang,M.Q. (2006). Computational prediction of methylation status in human genomic sequences. Proc. Natl. Acad. Sci. U. S. A. 103, 10713-10716.

de,I.C., X, Lois,S., Sanchez-Molina,S., and Martinez-Balbas,M.A. (2005). Do protein motifs read the histone code? Bioessays. 27, 164-175.

Delcuve,G.P., Rastegar,M., and Davie,J.R. (2009). Epigenetic control. J. Cell Physiol. 219, 243-250.

Deschler,B. and Lubbert,M. (2006). Acute myeloid leukemia: epidemiology and etiology. Cancer. 107, 2099-2107.

Dobrovic,A. and Kristensen,L.S. (2009). DNA methylation, epimutations and cancer predisposition. Int. J. Biochem. Cell Biol. 41, 34-39.

Dodge,J.E., List,A.F., and Futscher,B.W. (1998). Selective variegated methylation of the p15 CpG island in acute myeloid leukemia. Int. J. Cancer 78, 561-567.

References

Dodge,J.E., Munson,C., and List,A.F. (2001). KG-1 and KG-1a model the p15 CpG island methylation observed in acute myeloid leukemia patients. Leuk. Res. *25*, 917-925.

Douglas,D.B., Akiyama,Y., Carraway,H., Belinsky,S.A., Esteller,M., Gabrielson,E., Weitzman,S., Williams,T., Herman,J.G., and Baylin,S.B. (2004). Hypermethylation of a small CpGuanine-rich region correlates with loss of activator protein-2 alpha expression during progression of breast cancer. Cancer Research *64*, 1611-1620.

Ducasse,M. and Brown,M.A. (2006). Epigenetic aberrations and cancer. Mol. Cancer. *5:60.*, 60.

Ehrich,M., Nelson,M.R., Stanssens,P., Zabeau,M., Liloglou,T., Xinarianos,G., Cantor,C.R., Field,J.K., and van den,B.D. (2005). Quantitative high-throughput analysis of DNA methylation patterns by base-specific cleavage and mass spectrometry. Proc. Natl. Acad. Sci. U. S. A. *102*, 15785-15790.

Ehrlich,M. and Wang,R.Y.H. (1981). 5-Methylcytosine in Eukaryotic Dna. Science *212*, 1350-1357.

Estecio,M.R. and Issa,J.P. (2009). Tackling the methylome: recent methodological advances in genome-wide methylation profiling. Genome Med. *1*, 106.

Esteller,M. (2002). CpG island hypermethylation and tumor suppressor genes: a booming present, a brighter future. Oncogene. *21*, 5427-5440.

Esteller,M. (2005). Dormant hypermethylated tumour suppressor genes: questions and answers. J. Pathol. *205*, 172-180.

Esteller,M. (2006). CpG island methylation and histone modifications: biology and clinical significance. Ernst. Schering. Res. Found. Workshop. 115-126.

Esteller,M. (2007a). Cancer epigenomics: DNA methylomes and histone-modification maps. Nat. Rev. Genet. *8*, 286-298.

Esteller,M. (2007b). Epigenetic gene silencing in cancer: the DNA hypermethylome. Hum. Mol. Genet. *16 Spec No 1:R50-9.*, R50-R59.

Faniello,M.C., Bevilacqua,M.A., Condorelli,G., de Crombrugghe,B., Maity,S.N., Avvedimento,V.E., Cimino,F., and Costanzo,F. (1999). The B subunit of the CAAT-binding factor NFY binds the central segment of the co-activator p300. Journal of Biological Chemistry *274*, 7623-7626.

Farrell,W.E. (2005). Epigenetic mechanisms of tumorigenesis. Horm. Metab Res. *37*, 361-368.

Fazzari,M.J. and Greally,J.M. (2004). Epigenomics: Beyond CpG islands. Nature Reviews Genetics *5*, 446-455.

Feldman,N., Gerson,A., Fang,J., Li,E., Zhang,Y., Shinkai,Y., Cedar,H., and Bergman,Y. (2006). G9a-mediated irreversible epigenetic inactivation of Oct-3/4 during early embryogenesis. Nature Cell Biology *8*, 188-U55.

Feltus,F.A., Lee,E.K., Costello,J.F., Plass,C., and Vertino,P.M. (2003). Predicting aberrant CpG island methylation. Proc. Natl. Acad. Sci. U. S. A. *100*, 12253-12258.

Feltus,F.A., Lee,E.K., Costello,J.F., Plass,C., and Vertino,P.M. (2006). DNA motifs associated with aberrant CpG island methylation. Genomics. *87*, 572-579.

Feng,Q. and Zhang,Y. (2001). The MeCP1 complex represses transcription through preferential binding, remodeling, and deacetylating methylated nucleosomes. Genes & Development *15*, 827-832.

Figueiredo,L.M., Cross,G.A.M., and Janzen,C.J. (2009). Epigenetic regulation in African trypanosomes: a new kid on the block. Nature Reviews Microbiology *7*, 504-513.

Figueroa,M.E., Lugthart,S., Li,Y., Erpelinck-Verschueren,C., Deng,X., Christos,P.J., Schifano,E., Booth,J., van Putten,W., Skrabanek,L., Campagne,F., Mazumdar,M., Greally,J.M., Valk,P.J., Lowenberg,B., Delwel,R., and Melnick,A. (2010). DNA Methylation Signatures Identify Biologically Distinct Subtypes in Acute Myeloid Leukemia. Cancer Cell.

Figueroa,M.E., Wouters,B.J., Skrabanek,L., Glass,J., Li,Y.S., Erpelinck-Verschueren,C.A.J., Langerak,A.W., Lowenberg,B., Fazzari,M., Greally,J.M., Valk,P.J.M., Melnick,A., and Delwel,R. (2009). Genome-wide epigenetic analysis delineates a biologically distinct immature acute leukemia with myeloid/T-lymphoid features. Blood *113*, 2795-2804.

Flanagan,J.M. (2007). Host epigenetic modifications by oncogenic viruses. British Journal of Cancer *96*, 183-188.

Fraga,M.F., Ballestar,E., Montoya,G., Taysavang,P., Wade,P.A., and Esteller,M. (2003). The affinity of different MBD proteins for a specific methylated locus depends on their intrinsic binding properties. Nucleic Acids Res. *31*, 1765-1774.

Fraga,M.F., Ballestar,E., Villar-Garea,A., Boix-Chornet,M., Espada,J., Schotta,G., Bonaldi,T., Haydon,C., Ropero,S., Petrie,K., Iyer,N.G., Perez-Rosado,A., Calvo,E., Lopez,J.A., Cano,A., Calasanz,M.J., Colomer,D., Piris,M.A., Ahn,N., Imhof,A., Caldas,C., Jenuwein,T., and Esteller,M. (2005). Loss of acetylation at Lys16 and trimethylation at Lys20 of histone H4 is a common hallmark of human cancer. Nature Genetics *37*, 391-400.

Fraga,M.F. and Esteller,M. (2005). Towards the human cancer epigenome: a first draft of histone modifications. Cell Cycle. *4*, 1377-1381.

Fraga,M.F. and Esteller,M. (2007). Epigenetics and aging: the targets and the marks. Trends Genet. *23*, 413-418.

References

Frommer,M., McDonald,L.E., Millar,D.S., Collis,C.M., Watt,F., Grigg,G.W., Molloy,P.L., and Paul,C.L. (1992a). A genomic sequencing protocol that yields a positive display of 5-methylcytosine residues in individual DNA strands. Proc. Natl. Acad. Sci. U. S. A. *89*, 1827-1831.

Furukawa,Y., Sutheesophon,K., Wada,T., Nishimura,M., Saito,Y., Ishii,H., and Furukawa,Y. (2005). Methylation silencing of the Apaf-1 gene in acute leukemia. Molecular Cancer Research *3*, 325-334.

Galm,O., Herman,J.G., and Baylin,S.B. (2006). The fundamental role of epigenetics in hematopoietic malignancies. Blood Rev. *20*, 1-13.

Gardinergarden,M. and Frommer,M. (1987). Cpg Islands in Vertebrate Genomes. Journal of Molecular Biology *196*, 261-282.

Gebhard C, Benner C, Ehrich M, Schwarzfischer L, Schilling E, Klug M, Dietmaier W, Thiede C, Holler E, Andreesen R, and Rehli M (2010). General Transcription factor binding at CpG islands in normal cells correlates with resistance to de novo DNA methylation in cancer. Cancer Res.

Gebhard, C. Herstellung und Charakterisierung eines rekombinanten Methyl-CpG-bindenden Proteins. 2005. Naturwissenschaftliche Fakultät III, Universität Regensburg.

Ref Type: Thesis/Dissertation

Gebhard,C., Benner,C., Ehrich M, Schwarzfischer L, Schilling E, Klug M, Dietmaier W, Thiede C, Holler E, Andreesen R, and Rehli M (2010). General Transcription factor binding at CpG islands in normal cells correlates with resistance to de novo DNA methylation in cancer. Cancer Res.

Gebhard,C., Schwarzfischer,L., Pham,T.H., Andreesen,R., Mackensen,A., and Rehli,M. (2006a). Rapid and sensitive detection of CpG-methylation using methyl-binding (MB)-PCR. Nucleic Acids Res. *34*, e82.

Gebhard,C., Schwarzfischer,L., Pham,T.H., Schilling,E., Klug,M., Andreesen,R., and Rehli,M. (2006b). Genome-wide profiling of CpG methylation identifies novel targets of aberrant hypermethylation in myeloid leukemia. Cancer Res. *66*, 6118-6128.

Gehring,M., Reik,W., and Henikoff,S. (2009). DNA demethylation by DNA repair. Trends Genet. *25*, 82-90.

Goldberg,A.D., Allis,C.D., and Bernstein,E. (2007). Epigenetics: A landscape takes shape. Cell *128*, 635-638.

Grady,W.M. and Carethers,J.M. (2008). Genomic and epigenetic instability in colorectal cancer pathogenesis. Gastroenterology. *135*, 1079-1099.

Graw,R.G., Herzig,G.P., Eisel,R.J., and Perry,S. (1971). Leukocyte and Platelet Collection from Normal Donors with Continous Flow Blood Cell Separator. Transfusion *11*, 94-&.

Greene,F.L. (2007). Current TNM staging of colorectal cancer. Lancet Oncology *8*, 572-573.

Grubach,L., Juhl-Christensen,C., Rethmeier,A., Olesen,L.H., Aggerholm,A., Hokland,P., and Ostergaard,M. (2008). Gene expression profiling of Polycomb, Hox and Meis genes in patients with acute myeloid leukaemia. European Journal of Haematology *81*, 112-122.

Hackanson,B., Bennett,K.L., Brena,R.M., Jiang,J.M., Claus,R., Chen,S.S., Blagitko-Dorfs,N., Maharry,K., Whitman,S.P., Schmiittgen,T.D., Lubbert,M., Marcucci,G., Bloomfield,C.D., and Plass,C. (2008). Epigenetic modification of CCAAT/enhancer binding protein alpha expression in acute myeloid leukemia. Cancer Research *68*, 3142-3151.

Haehnel,V., Schwarzfischer,L., Fenton,M.J., and Rehli,M. (2002). Transcriptional regulation of the human toll-like receptor 2 gene in monocytes and macrophages. J. Immunol. *168*, 5629-5637.

Han,L., Lin,I.G., and Hsieh,C.L. (2001). Protein binding protects sites on stable episomes and in the chromosome from de novo methylation. Mol. Cell Biol. *21*, 3416-3424.

Hassan,A.H., Prochasson,P., Neely,K.E., Galasinski,S.C., Chandy,M., Carrozza,M.J., and Workman,J.L. (2002). Function and selectivity of bromodomains in anchoring chromatin-modifying complexes to promoter nucleosomes. Cell. *111*, 369-379.

Hatada,I., Fukasawa,M., Kimura,M., Morita,S., Yamada,K., Yoshikawa,T., Yamanaka,S., Endo,C., Sakurada,A., Sato,M., Kondo,T., Horii,A., Ushijima,T., and Sasaki,H. (2006). Genome-wide profiling of promoter methylation in human. Oncogene. *25*, 3059-3064.

Herman,J.G. and Baylin,S.B. (2003). Gene silencing in cancer in association with promoter hypermethylation. N. Engl. J. Med. *349*, 2042-2054.

Herman,J.G., Civin,C.I., Issa,J.P.J., Collector,M.I., Sharkis,S.J., and Baylin,S.B. (1997). Distinct patterns of inactivation of p15(INK4B) and p16(INK4A) characterize the major types of hematological malignancies. Cancer Research *57*, 837-841.

Herman,J.G., Jen,J., Merlo,A., and Baylin,S.B. (1996). Hypermethylation-associated inactivation indicates a tumor suppressor role for p15(INK4B1). Cancer Research *56*, 722-727.

Hirst,M. and Marra,M.A. (2009). Epigenetics and human disease. Int. J. Biochem. Cell Biol. *41*, 136-146.

References

Ho,K.L., McNae,I.W., Schmiedeberg,L., Klose,R.J., Bird,A.P., and Walkinshaw,M.D. (2008). MeCP2 binding to DNA depends upon hydration at methyl-CpG. Mol. Cell. *29*, 525-531.

Hui,J.Y., Stangl,K., Lane,W.S., and Bindereif,A. (2003). HnRNP L stimulates splicing of the eNOS gene by binding to variable-length CA repeats. Nature Structural Biology *10*, 33-37.

Imhof,A. (2006). Epigenetic regulators and histone modification. Brief. Funct. Genomic. Proteomic. *5*, 222-227.

Ionov,Y., Peinado,M.A., Malkhosyan,S., Shibata,D., and Perucho,M. (1993). Ubiquitous Somatic Mutations in Simple Repeated Sequences Reveal A New Mechanism for Colonic Carcinogenesis. Nature *363*, 558-561.

Irizarry,R.A., Ladd-Acosta,C., Carvalho,B., Wu,H., Brandenburg,S.A., Jeddeloh,J.A., Wen,B., and Feinberg,A.P. (2008). Comprehensive high-throughput arrays for relative methylation (CHARM). Genome Res. *18*, 780-790.

Irizarry,R.A., Ladd-Acosta,C., Wen,B., Wu,Z.J., Montano,C., Onyango,P., Cui,H.M., Gabo,K., Rongione,M., Webster,M., Ji,H., Potash,J.B., Sabunciyan,S., and Feinberg,A.P. (2009). The human colon cancer methylome shows similar hypo- and hypermethylation at conserved tissue-specific CpG island shores. Nature Genetics *41*, 178-186.

Issa,J.P. (2003). Age-related epigenetic changes and the immune system. Clin. Immunol. *109*, 103-108.

Issa,J.P. (2004). CpG island methylator phenotype in cancer. Nat. Rev. Cancer. *4*, 988-993.

Issa,J.P., Zehnbauer,B.A., Civin,C.I., Collector,M.I., Sharkis,S.J., Davidson,N.E., Kaufmann,S.H., and Baylin,S.B. (1996). The estrogen receptor CpG island is methylated in most hematopoietic neoplasms. Cancer Res. *56*, 973-977.

Izumi,H., Ohta,R., Nagatani,G., Ise,T., Nakayama,Y., Nomoto,M., and Kohno,K. (2003). p300/CBP-associated factor (P/CAF) interacts with nuclear respiratory factor-1 to regulate the UDP-N-acetyl-alpha-D-galactosamine: polypeptide N-acetylgalactosaminyltransferase-3 gene. Biochemical Journal *373*, 713-722.

Jabbour,E.J., Estey,E., and Kantarjian,H.M. (2006). Adult acute myeloid leukemia. Mayo Clin. Proc. *81*, 247-260.

Jacobson,R.H., Ladurner,A.G., King,D.S., and Tjian,R. (2000). Structure and function of a human TAF(II)250 double bromodomain module. Science *288*, 1422-1425.

Jin,B.L., Yao,B., Li,J.L., Fields,C.R., Delmas,A.L., Liu,C., and Robertson,K.D. (2009). DNMT1 and DNMT3B Modulate Distinct Polycomb-Mediated Histone Modifications in Colon Cancer. Cancer Research *69*, 7412-7421.

Johnson,W.D., Mei,B., and Cohn,Z.A. (1977). Separation, Long-Term Cultivation, and Maturation of Human Monocyte. Journal of Experimental Medicine *146*, 1613-1626.

Jones,P.A. and Baylin,S.B. (2002). The fundamental role of epigenetic events in cancer. Nat. Rev. Genet. *3*, 415-428.

Jones,P.A. and Baylin,S.B. (2007). The epigenomics of cancer. Cell. *128*, 683-692.

Jones,P.A., Wolkowicz,M.J., Rideout,W.M., III, Gonzales,F.A., Marziasz,C.M., Coetzee,G.A., and Tapscott,S.J. (1990). De novo methylation of the MyoD1 CpG island during the establishment of immortal cell lines. Proc. Natl. Acad. Sci. U. S. A. *87*, 6117-6121.

Jones,P.L., Veenstra,G.J., Wade,P.A., Vermaak,D., Kass,S.U., Landsberger,N., Strouboulis,J., and Wolffe,A.P. (1998). Methylated DNA and MeCP2 recruit histone deacetylase to repress transcription. Nat. Genet. *19*, 187-191.

Jost,J.P., Siegmann,M., Sun,L., and Leung,R. (1995). Mechanisms of DNA demethylation in chicken embryos. Purification and properties of a 5-methylcytosine-DNA glycosylase. J. Biol. Chem. *270*, 9734-9739.

Kafri,T., Ariel,M., Brandeis,M., Shemer,R., Urven,L., Mccarrey,J., Cedar,H., and Razin,A. (1992). Developmental Pattern of Gene-Specific Dna Methylation in the Mouse Embryo and Germ Line. Genes & Development *6*, 705-714.

Kelly,L.M., Englmeier,U., Lafon,I., Sieweke,M.H., and Graf,T. (2000a). MafB is an inducer of monocytic differentiation. EMBO J. *19*, 1987-1997.

Kelly,L.M., Englmeier,U., Lafon,I., Sieweke,M.H., and Graf,T. (2000b). MafB is an inducer of monocytic differentiation. Embo Journal *19*, 1987-1997.

Keshet,I., Schlesinger,Y., Farkash,S., Rand,E., Hecht,M., Segal,E., Pikarski,E., Young,R.A., Niveleau,A., Cedar,H., and Simon,I. (2006). Evidence for an instructive mechanism of de novo methylation in cancer cells. Nat. Genet. *38*, 149-153.

Kikuchi,T., Toyota,M., Itoh,F., Suzuki,H., Obata,T., Yamamoto,H., Kakiuchi,H., Kusano,M., Issa,J.P.J., Tokino,T., and Imai,K. (2002). Inactivation of p57KIP2 by regional promoter hypermethylation and histone deacetylation in human tumors. Oncogene *21*, 2741-2749.

Kimura,H. and Shiota,K. (2003). Methyl-CpG-binding protein, MeCP2, is a target molecule for maintenance DNA methyltransferase, Dnmt1. Journal of Biological Chemistry *278*, 4806-4812.

Klose,R.J. and Bird,A.P. (2006). Genomic DNA methylation: the mark and its mediators. Trends Biochem. Sci. *31*, 89-97.

Klose,R.J., Sarraf,S.A., Schmiedeberg,L., McDermott,S.M., Stancheva,I., and Bird,A.P. (2005). DNA binding selectivity of MeCP2 due to a requirement for A/T sequences adjacent to methyl-CpG. Mol. Cell. *19*, 667-678.

Kornberg,R.D. (1974). Chromatin structure: a repeating unit of histones and DNA. Science. *184*, 868-871.

References

Kornberg,R.D. and Lorch,Y. (1999). Twenty-five years of the nucleosome, fundamental particle of the eukaryote chromosome. Cell. *98*, 285-294.

Kouzarides,T. (2007). Chromatin modifications and their function. Cell. *128*, 693-705.

Kress,C., Thomassin,H., and Grange,T. (2006). Active cytosine demethylation triggered by a nuclear receptor involves DNA strand breaks. Proc. Natl. Acad. Sci. U. S. A. *103*, 11112-11117.

Kroeger,H., Jelinek,J., Estecio,M.R.H., He,R., Kondo,K., Chung,W., Zhang,L., Shen,L.L., Kantarjian,H.M., Bueso-Ramos,C.E., and Issa,J.P.J. (2008). Aberrant CpG island methylation in acute myeloid leukemia is accentuated at relapse. Blood *112*, 1366-1373.

Ku,M., Koche,R.P., Rheinbay,E., Mendenhall,E.M., Endoh,M., Mikkelsen,T.S., Presser,A., Nusbaum,C., Xie,X.H., Chi,A.S., Adli,M., Kasif,S., Ptaszek,L.M., Cowan,C.A., Lander,E.S., Koseki,H., and Bernstein,B.E. (2008). Genomewide Analysis of PRC1 and PRC2 Occupancy Identifies Two Classes of Bivalent Domains. Plos Genetics *4*.

Kurkjian,C., Kummar,S., and Murgo,A.J. (2008). DNA Methylation: Its Role in Cancer Development and Therapy. Current Problems in Cancer *32*, 185-235.

Laird,P.W. (2005). Cancer epigenetics. Hum. Mol. Genet. *14 Spec No 1:R65-76.*, R65-R76.

Larsson,J. and Karlsson,S. (2005). The role of Smad signaling in hematopoiesis. Oncogene *24*, 5676-5692.

Lemon,B. and Tjian,R. (2000). Orchestrated response: a symphony of transcription factors for gene control. Genes & Development *14*, 2551-2569.

Li,E., Bestor,T.H., and Jaenisch,R. (1992). Targeted mutation of the DNA methyltransferase gene results in embryonic lethality. Cell. *69*, 915-926.

Lin,I.G. and Hsieh,C.L. (2001). Chromosomal DNA demethylation specified by protein binding. EMBO Rep. *2*, 108-112.

Lindroth,A.M., Park,Y.J., Mclean,C.M., Dokshin,G.A., Persson,J.M., Herman,H., Pasini,D., Miro,X., Donohoe,M.E., Lee,J.T., Helin,K., and Soloway,P.D. (2008). Antagonism between DNA and H3K27 Methylation at the Imprinted Rasgrf1 Locus. Plos Genetics *4*.

Lister,R. and Ecker,J.R. (2009). Finding the fifth base: Genome-wide sequencing of cytosine methylation. Genome Research *19*, 959-966.

Lister,R., O'Malley,R.C., Tonti-Filippini,J., Gregory,B.D., Berry,C.C., Millar,A.H., and Ecker,J.R. (2008). Highly integrated single-base resolution maps of the epigenome in Arabidopsis. Cell *133*, 523-536.

Liu,Y., Taverna,S.D., Muratore,T.L., Shabanowitz,J., Hunt,D.F., and Allis,C.D. (2007). RNAi-dependent H3K27 methylation is required for heterochromatin formation and DNA elimination in Tetrahymena. Genes & Development *21*, 1530-1545.

Liu,Z.J., Zhang,X.B., Zhang,Y., and Yang,X. (2004). Progesterone receptor gene inactivation and CpG island hypermethylation in human leukemia cancer cells. Febs Letters *567*, 327-332.

Lowenberg,B. (2008). Acute myeloid leukemia: the challenge of capturing disease variety. Hematology. Am. Soc. Hematol. Educ. Program. 1-11.

Loyola,A. and Almouzni,G. (2004). Histone chaperones, a supporting role in the limelight. Biochimica et Biophysica Acta-Gene Structure and Expression *1677*, 3-11.

Lund,A.H. and van Lohuizen,M. (2004). Epigenetics and cancer. Genes & Development *18*, 2315-2335.

Macleod,D., Charlton,J., Mullins,J., and Bird,A.P. (1994). Sp1 sites in the mouse aprt gene promoter are required to prevent methylation of the CpG island. Genes Dev. *8*, 2282-2292.

Martin-Subero,J.I., Ammerpohl,O., Bibikova,M., Wickham-Garcia,E., Agirre,X., Alvarez,S., Bruggemann,M., Bug,S., Calasanz,M.J., Deckert,M., Dreyling,M., Du,M.Q., Durig,J., Dyer,M.J., Fan,J.B., Gesk,S., Hansmann,M.L., Harder,L., Hartmann,E., Klapper,W., Kuppers,R., Montesinos-Rongen,M., Nagel,I., Pott,C., Richter,J., Roman-Gomez,J., Seifert,M., Stein,H., Suela,J., Trumper,L., Vater,I., Prosper,F., Haferlach,C., Cruz,C.J., and Siebert,R. (2009). A comprehensive microarray-based DNA methylation study of 367 hematological neoplasms. PLoS One. *4*, e6986.

Meissner,A., Mikkelsen,T.S., Gu,H., Wernig,M., Hanna,J., Sivachenko,A., Zhang,X., Bernstein,B.E., Nusbaum,C., Jaffe,D.B., Gnirke,A., Jaenisch,R., and Lander,E.S. (2008). Genome-scale DNA methylation maps of pluripotent and differentiated cells. Nature. *454*, 766-770.

Metivier,R., Gallais,R., Tiffoche,C., Le Peron,C., Jurkowska,R.Z., Carmouche,R.P., Ibberson,D., Barath,P., Demay,F., Reid,G., Benes,V., Jeltsch,A., Gannon,F., and Salbert,G. (2008). Cyclical DNA methylation of a transcriptionally active promoter. Nature. *452*, 45-50.

Moazed,D. (2007). RNAi and gene silencing in heterochromatin. Febs Journal *274*, 19.

Moazed,D., Verdel,A., Motamedi,M.R., Colmenares,S., Gerber,S.A., and Gygi,S.P. (2005). RNAi complexes, noncoding RNAs, and heterochromatin assembly. Faseb Journal *19*, A1720.

Mohn,F. and Schubeler,D. (2009). Genetics and epigenetics: stability and plasticity during cellular differentiation. Trends Genet. *25*, 129-136.

References

Monk,M., Boubelik,M., and Lehnert,S. (1987). Temporal and Regional Changes in Dna Methylation in the Embryonic, Extraembryonic and Germ-Cell Lineages During Mouse Embryo Development. Development *99*, 371-382.

Muller,J., Hart,C.M., Francis,N.J., Vargas,M.L., Sengupta,A., Wild,B., Miller,E.L., O'Connor,M.B., Kingston,R.E., and Simon,J.A. (2002). Histone methyltransferase activity of a Drosophila polycomb group repressor complex. Cell *111*, 197-208.

Mullis,K., Faloona,F., Scharf,S., Saiki,R., Horn,G., and Erlich,H. (1986). Specific Enzymatic Amplification of Dna Invitro - the Polymerase Chain-Reaction. Cold Spring Harbor Symposia on Quantitative Biology *51*, 263-273.

Murai,M., Toyota,M., Satoh,A., Suzuki,H., Akino,K., Mita,H., Sasaki,Y., Ishida,T., Shen,L., Garcia-Manero,G., Issa,J.P.J., Hinoda,Y., Tokino,T., and Imai,K. (2005). Aberrant DNA methylation associated with silencing BNIP3 gene expression in haematopoietic tumours. British Journal of Cancer *92*, 1165-1172.

Mutskov,V.J., Farrell,C.M., Wade,P.A., Wolffe,A.P., and Felsenfeld,G. (2002). The barrier function of an insulator couples high histone acetylation levels with specific protection of promoter DNA from methylation. Genes & Development *16*, 1540-1554.

Nakayama,T., Nishioka,K., Dong,Y.X., Shimojima,T., and Hirose,S. (2007). Drosophila GAGA factor directs histone H3.3 replacement that prevents the heterochromatin spreading. Genes Dev. *21*, 552-561.

Nan,X., Ng,H.H., Johnson,C.A., Laherty,C.D., Turner,B.M., Eisenman,R.N., and Bird,A. (1998). Transcriptional repression by the methyl-CpG-binding protein MeCP2 involves a histone deacetylase complex. Nature. *393*, 386-389.

Ng,H.H. and Bird,A. (1999). DNA methylation and chromatin modification. Curr. Opin. Genet. Dev. *9*, 158-163.

Ohgane,J., Yagi,S., and Shiota,K. (2008). Epigenetics: the DNA methylation profile of tissue-dependent and differentially methylated regions in cells. Placenta. *29 Suppl A:S29-35. Epub;%2007 Nov 26.*, S29-S35.

Ohm,J.E., McGarvey,K.M., Yu,X., Cheng,L.Z., Schuebel,K.E., Cope,L., Mohammad,H.P., Chen,W., Daniel,V.C., Yu,W., Berman,D.M., Jenuwein,T., Pruitt,K., Sharkis,S.J., Watkins,D.N., Herman,J.G., and Baylin,S.B. (2007). A stem cell-like chromatin pattern may predispose tumor suppressor genes to DNA hypermethylation and heritable silencing. Nature Genetics *39*, 237-242.

Okano,M., Bell,D.W., Haber,D.A., and Li,E. (1999). DNA methyltransferases Dnmt3a and Dnmt3b are essential for de novo methylation and mammalian development. Cell. *99*, 247-257.

Orkin,S.H. (2000). Diversification of haematopoietic stem cells to specific lineages. Nature Reviews Genetics *1*, 57-64.

Paz,M.F., Fraga,M.F., Avila,S., Guo,M., Pollan,M., Herman,J.G., and Esteller,M. (2003). A systematic profile of DNA methylation in human cancer cell lines. Cancer Res. *63*, 1114-1121.

Peters,A.H. and Schubeler,D. (2005). Methylation of histones: playing memory with DNA. Curr. Opin. Cell Biol. *17*, 230-238.

Pfeifer,G.P. and Besaratinia,A. (2009). Mutational spectra of human cancer. Hum. Genet. *125*, 493-506.

Pietersen,A.M. and van Lohuizen,M. (2008). Stem cell regulation by polycomb repressors: postponing commitment. Current Opinion in Cell Biology *20*, 201-207.

Plass,C. (2002). Cancer epigenomics. Hum. Mol. Genet. *11*, 2479-2488.

Plass,C., Oakes,C., Blum,W., and Marcucci,G. (2008). Epigenetics in acute myeloid leukemia. Semin. Oncol. *35*, 378-387.

Plass,C. and Soloway,P.D. (2002). DNA methylation, imprinting and cancer. Eur. J. Hum. Genet. *10*, 6-16.

Razin,A. (1998). CpG methylation, chromatin structure and gene silencing-a three-way connection. EMBO J. *17*, 4905-4908.

Reid,G., Gallais,R., and Metivier,R. (2009). Marking time: The dynamic role of chromatin and covalent modification in transcription. International Journal of Biochemistry & Cell Biology *41*, 155-163.

Reik,W., Dean,W., and Walter,J. (2001). Epigenetic reprogramming in mammalian development. Science. *293*, 1089-1093.

Rethmeier,A., Aggerholm,A., Olesen,L.H., Juhl-Christensen,C., Nyvold,C.G., Guldberg,P., and Hokland,P. (2006). Promoter hypermethylation of the retinoic acid receptor beta 2 gene is frequent in acute myeloid leukaemia and associated with the presence of CBF beta-MYH11 fusion transcripts. British Journal of Haematology *133*, 276-283.

Rice,J.C. and Allis,C.D. (2001). Histone methylation versus histone acetylation: new insights into epigenetic regulation. Curr. Opin. Cell Biol. *13*, 263-273.

Ringrose,L. and Paro,R. (2007). Polycomb/Trithorax response elements and epigenetic memory of cell identity. Development. *134*, 223-232.

Robertson,K.D. (2002). DNA methylation and chromatin - unraveling the tangled web. Oncogene. *21*, 5361-5379.

Roman-Gomez,J., Castillejo,J.A., Jimenez,A., Cervantes,F., Boque,C., Hermosin,L., Leon,A., Granena,A., Colomer,D., Heiniger,A., and Torres,A. (2003). Cadherin-13, a mediator of calcium-dependent cell-cell adhesion, is silenced by methylation in chronic myeloid leukemia and correlates with pretreatment risk profile and cytogenetic response to interferon alfa. Journal of Clinical Oncology 21, 1472-1479.

Rozenberg,J.M., Shlyakhtenko,A., Glass,K., Rishi,V., Myakishev,M.V., FitzGerald,P.C., and Vinson,C. (2008). All and only CpG containing sequences are enriched in promoters abundantly bound by RNA polymerase II in multiple tissues. Bmc Genomics 9.

Rush,L.J. and Plass,C. (2002). Restriction landmark genomic scanning for DNA methylation in cancer: past, present, and future applications. Anal. Biochem. 307, 191-201.

Sanderson,R.J., Shepperdson,F.T., Vatter,A.E., and Talmage,D.W. (1977). Isolation and Enumeration of Peripheral-Blood Monocytes. Journal of Immunology 118, 1409-1414.

Sarraf,S.A. and Stancheva,I. (2004). Methyl-CpG binding protein MBD1 couples histone H3 methylation at lysine 9 by SETDB1 to DNA replication and chromatin assembly. Mol. Cell. 15, 595-605.

Sato,K., Tamura,G., Tsuchiya,T., Endoh,Y., Usuba,O., Kimura,W., and Motoyama,T. (2001). Frequent loss of expression without sequence mutations of the DCC gene in primary gastric cancer. British Journal of Cancer 85, 199-203.

Sato,N., Fukui,T., Taniguchi,T., Yokoyama,T., Kondo,M., Nagasaka,T., Goto,Y., Gao,W.T., Ueda,Y., Yokoi,K., Minna,J.D., Osada,H., Kondo,Y., and Sekido,Y. (2007). RhoB is frequently downregulated in non-small-cell lung cancer and resides in the 2p24 homozygous deletion region of a lung cancer cell line. International Journal of Cancer 120, 543-551.

Scardocci,A., Guidi,F., D'Alo,F., Gumiero,D., Fabiani,E., DiRuscio,A., Martini,M., Larocca,L.M., Zollino,M., Hohaus,S., Leone,G., and Voso,M.T. (2006). Reduced BRCA1 expression due to promoter hypermethylation in therapy-related acute myeloid leukaemia. British Journal of Cancer 95, 1108-1113.

Schilling,E., El Chartouni,C., and Rehli,M. (2009). Allele-specific DNA methylation in mouse strains is mainly determined by cis-acting sequences. Genome Research 19, 2028-2035.

Schilling,E. and Rehli,M. (2007). Global, comparative analysis of tissue-specific promoter CpG methylation. Genomics. 90, 314-323.

Schlesinger,Y., Straussman,R., Keshet,I., Farkash,S., Hecht,M., Zimmerman,J., Eden,E., Yakhini,Z., Ben Shushan,E., Reubinoff,B.E., Bergman,Y., Simon,I., and Cedar,H. (2007). Polycomb-mediated methylation on Lys27 of histone H3 pre-marks genes for de novo methylation in cancer. Nat. Genet. 39, 232-236.

Schmidl,C., Klug,M., Boedl,T.J., Andreesen,R., Hoffmann,P., Edinger,M., and Rehli,M. (2009). Lineage-specific DNA methylation in T cells correlates with histone methylation and enhancer activity. Genome Res. 19, 1165-1174.

Schwab,J. and Illges,H. (2001). Regulation of CD21 expression by DNA methylation and histone deacetylation. International Immunology 13, 705-711.

Shen,L.L., Toyota,M., Kondo,Y., Obata,T., Daniel,S., Pierce,S., Imai,K., Kantarjian,H.M., Issa,J.P.J., and Garcia-Manero,G. (2003). Aberrant DNA methylation of p57KIP2 identifies a cell-cycle regulatory pathway with prognostic impact in adult acute lymphocytic leukemia. Blood 101, 4131-4136.

Shipley,J.L. and Butera,J.N. (2009). Acute myelogenous leukemia. Experimental Hematology 37, 649-658.

Shiraishi,M., Chuu,Y.H., and Sekiya,T. (1999). Isolation of DNA fragments associated with methylated CpG islands in human adenocarcinomas of the lung using a methylated DNA binding column and denaturing gradient gel electrophoresis. Proc. Natl. Acad. Sci. U. S. A 96, 2913-2918.

Siegfried,Z., Eden,S., Mendelsohn,M., Feng,X., Tsuberi,B.Z., and Cedar,H. (1999). DNA methylation represses transcription in vivo. Nature Genetics 22, 203-206.

Singal,R. and Ginder,G.D. (1999). DNA methylation. Blood 93, 4059-4070.

Smiraglia,D.J., Rush,L.J., Fruhwald,M.C., Dai,Z.Y., Held,W.A., Costello,J.F., Lang,J.C., Eng,C., Li,B., Wright,F.A., Caligiuri,M.A., and Plass,C. (2001). Excessive CpG island hypermethylation in cancer cell lines versus primary human malignancies. Human Molecular Genetics 10, 1413-1419.

Smith,R.J., Dean,W., Konfortova,G., and Kelsey,G. (2003). Identification of novel imprinted genes in a genome-wide screen for maternal methylation. Genome Res. 13, 558-569.

Steffen,B., Muller-Tidow,C., Schwable,J., Berdel,W.E., and Serve,H. (2005). The molecular pathogenesis of acute myeloid leukemia. Crit Rev. Oncol. Hematol. 56, 195-221.

Stone,R.M., O'Donnell,M.R., and Sekeres,M.A. (2004). Acute myeloid leukemia. Hematology. Am. Soc. Hematol. Educ. Program. 98-117.

Strahl,B.D. and Allis,C.D. (2000). The language of covalent histone modifications. Nature. 403, 41-45.

Straussman,R., Nejman,D., Roberts,D., Steinfeld,I., Blum,B., Benvenisty,N., Simon,I., Yakhini,Z., and Cedar,H. (2009). Developmental programming of CpG island methylation profiles in the human genome. Nat. Struct. Mol. Biol. 16, 564-571.

References

Suzuki,H., Gabrielson,E., Chen,W., Anbazhagan,R., van Engeland,M., Weijenberg,M.P., Herman,J.G., and Baylin,S.B. (2002). A genomic screen for genes upregulated by demethylation and histone deacetylase inhibition in human colorectal cancer. Nat. Genet. *31*, 141-149.

Suzuki,M., Yamada,T., Kihara-Negishi,F., Sakurai,T., Hara,E., Tenen,D.G., Hozumi,N., and Oikawa,T. (2006). Site-specific DNA methylation by a complex of PU.1 and Dnmt3a/b. Oncogene *25*, 2477-2488.

Suzuki,M.M. and Bird,A. (2008). DNA methylation landscapes: provocative insights from epigenomics. Nat. Rev. Genet. *9*, 465-476.

Tagoh,H., Melnik,S., Lefevre,P., Chong,S., Riggs,A.D., and Bonifer,C. (2004). Dynamic reorganization of chromatin structure and selective DNA demethylation prior to stable enhancer complex formation during differentiation of primary hematopoietic cells in vitro. Blood. *103*, 2950-2955.

Takai,D. and Jones,P.A. (2002). Comprehensive analysis of CpG islands in human chromosomes 21 and 22. Proceedings of the National Academy of Sciences of the United States of America *99*, 3740-3745.

Takeshima,H., Yamashita,S., Shimazu,T., Niwa,T., and Ushijima,T. (2009). The presence of RNA polymerase II, active or stalled, predicts epigenetic fate of promoter CpG islands. Genome Research *19*, 1974-1982.

Taverna,S.D., Li,H., Ruthenburg,A.J., Allis,C.D., and Patel,D.J. (2007). How chromatin-binding modules interpret histone modifications: lessons from professional pocket pickers. Nat. Struct. Mol. Biol. *14*, 1025-1040.

Thomassin,H., Flavin,M., Espinas,M.L., and Grange,T. (2001). Glucocorticoid-induced DNA demethylation and gene memory during development. EMBO J. *20*, 1974-1983.

Toyota,M., Kopecky,K.J., Toyota,M.O., Jair,K.W., Willman,C.L., and Issa,J.P.J. (2001). Methylation profiling in acute myeloid leukemia. Blood *97*, 2823-2829.

Turker,M.S. (2002). Gene silencing in mammalian cells and the spread of DNA methylation. Oncogene. *21*, 5388-5393.

Turner,B.M. (2007). Defining an epigenetic code. Nature Cell Biology *9*, 2-6.

Valk,P.J.M., Verhaak,R.G.W., Beijen,M.A., Erpelinck,C.A.J., Doorn-Khosrovani,S.B.V., Boer,J.M., Beverloo,H.B., Moorhouse,M.J., van der Spek,P.J., Lowenberg,B., and Delwel,R. (2004). Prognostically useful gene-expression profiles in acute myeloid leukemia. New England Journal of Medicine *350*, 1617-1628.

Vardiman,J.W., Harris,N.L., and Brunning,R.D. (2002). The World Health Organization (WHO) classification of the myeloid neoplasms. Blood *100*, 2292-2302.

Verhaak,R.G.W., Wouters,B.J., Erpelinck,C.A.J., Abbas,S., Beverloo,H.B., Lugthart,S., Lowenberg,B., Delwel,R., and Valk,P.J.M. (2009). Prediction of molecular subtypes in acute myeloid leukemia based on gene expression profiling. Haematologica-the Hematology Journal *94*, 131-134.

Vire,E., Brenner,C., Deplus,R., Blanchon,L., Fraga,M., Didelot,C., Morey,L., Van Eynde,A., Bernard,D., Vanderwinden,J.M., Bollen,M., Esteller,M., Di Croce,L., de Launoit,Y., and Fuks,F. (2006). The Polycomb group protein EZH2 directly controls DNA methylation. Nature *439*, 871-874.

Wade,P.A., Gegonne,A., Jones,P.L., Ballestar,E., Aubry,F., and Wolffe,A.P. (1999). Mi-2 complex couples DNA methylation to chromatin remodelling and histone deacetylation. Nature Genetics *23*, 62-66.

Walsh,C.P. and Bestor,T.H. (1999). Cytosine methylation and mammalian development. Genes Dev. *13*, 26-34.

Warnecke,P.M. and Clark,S.J. (1999). DNA methylation profile of the mouse skeletal alpha-actin promoter during development and differentiation. Molecular and Cellular Biology *19*, 164-172.

Weber,M., Davies,J.J., Wittig,D., Oakeley,E.J., Haase,M., Lam,W.L., and Schubeler,D. (2005). Chromosome-wide and promoter-specific analyses identify sites of differential DNA methylation in normal and transformed human cells. Nat. Genet. *37*, 853-862.

Weber,M., Hellmann,I., Stadler,M.B., Ramos,L., Paabo,S., Rebhan,M., and Schubeler,D. (2007). Distribution, silencing potential and evolutionary impact of promoter DNA methylation in the human genome. Nat. Genet. *39*, 457-466.

Weissmann,F. and Lyko,F. (2003). Cooperative interactions between epigenetic modifications and their function in the regulation of chromosome architecture. Bioessays. *25*, 792-797.

Whitman,S.P., Hackanson,B., Liyanarachchi,S., Liu,S., Rush,L.J., Maharry,K., Margeson,D., Davuluri,R., Wen,J., Witte,T., Yu,L., Liu,C., Bloomfield,C.D., Marcucci,G., Plass,C., and Caligiuri,M.A. (2008). DNA hypermethylation and epigenetic silencing of the tumor suppressor gene, SLC5A8, in acute myeloid leukemia with the MLL partial tandem duplication. Blood *112*, 2013-2016.

Widschwendter,M., Fiegl,H., Egle,D., Mueller-Holzner,E., Spizzo,G., Marth,C., Weisenberger,D.J., Campan,M., Young,J., Jacobs,I., and Laird,P.W. (2007). Epigenetic stem cell signature in cancer. Nat. Genet. *39*, 157-158.

Wikipedia contributors. Haematopoiesis. In Wikipedia, The Free Encyclopedia . 2010. Wikipedia, The Free Encyclopedia. 19-1-2010.

Ref Type: Electronic Citation

Wolffe,A.P., Jones,P.L., and Wade,P.A. (1999). DNA demethylation. Proc. Natl. Acad. Sci. U. S. A *96*, 5894-5896.

References

Wong,J.J., Hawkins,N.J., and Ward,R.L. (2007). Colorectal cancer: a model for epigenetic tumorigenesis. Gut. *56*, 140-148.

Wouters,B.J., Jorda,M.A., Keeshan,K., Louwers,I., Erpelinck-Verschueren,C.A.J., Tielemans,D., Langerak,A.W., He,Y.P., Yashiro-Ohtani,Y., Zhang,P., Hetherington,C.J., Verhaak,R.G.W., Valk,P.J.M., Lowenberg,B., Tenen,D.G., Pear,W.S., and Delwel,R. (2007). Distinct gene expression profiles of acute myeloid/T-lymphoid leukemia with silenced CEBPA and mutations in NOTCH1. Blood *110*, 3706-3714.

Xie,X., Lu,J., Kulbokas,E.J., Golub,T.R., Mootha,V., Lindblad-Toh,K., Lander,E.S., and Kellis,M. (2005a). Systematic discovery of regulatory motifs in human promoters and 3' UTRs by comparison of several mammals. Nature *434*, 338-345.

Xie,X., Lu,J., Kulbokas,E.J., Golub,T.R., Mootha,V., Lindblad-Toh,K., Lander,E.S., and Kellis,M. (2005b). Systematic discovery of regulatory motifs in human promoters and 3' UTRs by comparison of several mammals. Nature. *434*, 338-345.

Yasunaga,J., Taniguchi,Y., Nosaka,K., Yoshida,M., Satou,Y., Sakai,T., Mitsuya,H., and Matsuoka,M. (2004). Identification of aberrantly methylated genes in association with adult T-cell leukemia. Cancer Research *64*, 6002-6009.

Youssef,E.M., Chen,X.Q., Higuchi,E., Kondo,Y., Garcia-Manero,G., Lotan,R., and Issa,J.P.J. (2004). Hypermethylation and silencing of the putative tumor suppressor Tazarotene-induced gene 1 in human cancers. Cancer Research *64*, 2411-2417.

Yu,L., Liu,C.H., Vandeusen,J., Becknell,B., Dai,Z.Y., Wu,Y.Z., Raval,A., Liu,T.H., Ding,W., Mao,C., Liu,S.J., Smith,L.T., Lee,S., Rassenti,L., Marcucci,G., Byrd,J., Caligiuri,M.A., and Plass,C. (2005). Global assessment of promoter methylation in a mouse model of cancer identifies ID4 as a putative tumor-suppressor gene in human leukemia. Nature Genetics *37*, 265-274.

Yuan,B.Z., Durkin,M.E., and Popescu,N.C. (2003). Promoter hypermethylation of DLC-1, a candidate tumor suppressor gene, in several common human cancers. Cancer Genetics and Cytogenetics *140*, 113-117.

Zeschnigk,M., Schmitz,B., Dittrich,B., Buiting,K., Horsthemke,B., and Doerfler,W. (1997). Imprinted segments in the human genome: Different DNA methylation patterns in the Prader-Willi/Angelman syndrome region as determined by the genomic sequencing method. Human Molecular Genetics *6*, 387-395.

Zhu,B., Zheng,Y., Hess,D., Angliker,H., Schwarz,S., Siegmann,M., Thiry,S., and Jost,J.P. (2000). 5-methylcytosine-DNA glycosylase activity is present in a cloned G/T mismatch DNA glycosylase associated with the chicken embryo DNA demethylation complex. Proceedings of the National Academy of Sciences of the United States of America *97*, 5135-5139.

10 Abbreviations

AML	Acute myeloid leukemia
AS	Antisense
5mC	5-methylcytosine
bp	Base pair
BS	Bisulfite
BSA	Bovine serum albumin
°C	Degree Celsius
cDNA	Complementary DNA
CGI	CpG island
ChIP	Chromatin immunoprecipitation
CIAP	Calf intestinal alkaline phosphatase
CpG	Cytosine-guanine dinucleotide
dd	Double distilled
DEPC	Diethyl pyrocarbonate
DMEM	Dulbecco's Modified Eagle Medium
DMR	Differential methylated region
DMSO	Dimethyl sulfoyde
DNMT	DNA methyltransferase
dNTP	Deoxiribonucleotide triphosphate
ECL	Enhanced chemiluminescence
EDTA	Ethylenediaminetetraacetic acid
ES cell	Embryonic stem cell
EtOH	Ethanol
FACS	Fluorescence activated cell sorting
FCS	Fetal Calf Serum
gDNA	Genomic DNA
GO	Gene ontology
H	Hour
HELP	*Hpa* II tiny fragment Enrichment by LM-PCR
HSC	Hematopoietic stem cell
H3K4me1	Histone 3 lysine 4 monomethylation
H3K4me2	Histone 3 lysine 4 dimethylation
H3K4me3	Histone 3 lysine 4 trimethylation
HAT	Histone acetyltransferase
HDAC	Histone deacetylase

Abbreviations

HMT	Histone methyltransferase
IP	Immunoprecipitation
LM-PCR	Ligation-mediated polymerase chain reaction
MALDI-TOF MS	Matrix-assisted laser desorption/ionization time-of-flight mass spectrometry
MBD	Methyl-CpG binding domain
MCIp	Methyl-CpG immunoprecipitation
MeCP2	Methyl-CpG binding Protein 2
MeDIP	Methylated DNA immunoprecipitation
Min	Minute
MO	Monocyte
MOPS	3-(N-Morpholino) propanesulfonic acid
mRNA	Messenger RNA
MSP	Methyl-specific PCR
MvA	Signal log ratio vs. average log intensity
NaOAc	Sodium acetate
NK cell	Natural killer cell
NP-40	Nonidet P-40
O/N	Overnight
PB-MNCs	Peripheral blood mononuclear cells
PBS	Phosphate buffered saline
PEG	Polyethyleneglycol
PCR	Polymerase chain reaction
qPCR	Quantitative PCR
RLGS	Restriction landmark genomic scanning
rpm	Rounds per minute
RT	Room temperature
RT-qPCR	Quantitative reverse transcription PCR
s	Second
S	Sense
SAM	S-adenosylmethionine
SAP	Shrimp alkaline phosphatase
SD	Standard deviation
SDS	Sodium dodecyl sulfate
SNP	Single nucleotide polymorphism
TAE	Tris acetate /EDTA electrophoresis buffer
TE	Tris-EDTA
TEMED	N,N,N',N',-Tetramethylenediamine

Abbreviations

TSS	Transcription start site
UCSC	University of California, Santa Cruz
X-gal	5-Bromo-4-chloro-3-indoyl-β-D-galactopyranosid

11 Publications

Rapid and sensitive detection of CpG-methylation using methyl-binding (MB)-PCR
Gebhard C, Schwarzfischer L, Pham TH, Andreesen R, Mackensen A, Rehli M.
Nucleic Acids Res. 2006 Jul 5;34(11):e82.

Genome-wide profiling of CpG methylation identifies novel targets of aberrant hypermethylation in myeloid leukemia
Gebhard C, Schwarzfischer L, Pham TH, Schilling E, Klug M, Andreesen R, Rehli M
Cancer Res. 2006 Jun 15;66(12):6118-28

General Transcription factor binding at CpG islands in normal cells correlates with resistance to de novo DNA methylation in cancer
Gebhard C, Benner C, Ehrich M, Schwarzfischer L, Schilling E, Klug M, Dietmaier W, Thiede C, Holler E, Andreesen R, Rehli M
Cancer Res. 2010 Feb 70(4)

Active DNA demethylation in human postmitotic cells correlates with activating histone modifications, but not transcription levels
Klug M, Heinz S, Gebhard C, Schwarzfischer L, Krause S, Andreesen R, Rehli M
(submitted)

12 Acknowledgement

Für die Ermöglichung dieser Doktorarbeit und seine großzügige Unterstützung während dieser Zeit möchte ich mich sehr herzlich bei Prof. Dr. Reinhard Andreesen bedanken.

Bei Prof. Dr. Stephan Schneuwly bedanke ich mich sehr für die Betreuung und Begutachtung dieser Arbeit.

Mein besonderer Dank gilt Prof. Dr. Michael Rehli für die ausgezeichnete Betreuung während all der Jahre, sein großes Interesse und seine endlose Geduld. Seine Tür stand immer offen für Diskussionen, Anregungen und Fragen. Danke für den San Diego-Aufenthalt und die Teilnahme an vielen Kongressen.

Maja, was hätte ich ohne dich gemacht...gut, dass du zu uns nach Regensburg gekommen bist. Danke für die lustige Zeit, sowohl im Labor, als auch außerhalb (VGTs...). Für deine Hilfsbereitschaft und Unterstützung zu jeder Zeit, die ganzen Aktionen, die wir zusammen durchgezogen haben, und dass du mir geholfen hast, ständig die Schokoschubladen zu leeren ☺

Vielen Dank Lucia, für deine tatkräftige Unterstützung zu jeder Zeit. Auf jede Frage hattest du eine passende Antwort. Und ohne deine Ordnung hätten wir so manche Plasmide oder Sequenzierprimer nicht mehr gefunden ☺

Ein besonderer Dank gilt Dagmar für ihren Einsatz und ihre Unterstützung, v.a. mit dem Massenspektrometer (du hast auch in nervenaufreibenden Situationen Ruhe bewahrt ☺).

Vielen Dank an meine beiden Schatzis Moni und Carol! Schade, dass ihr nicht mehr in unserer Arbeitsgruppe seid! Vielen Dank, dass ihr in jeder Lebenslage für mich da seid. Moni, ohne dich wäre Krankengymnastik nur halb so lustig gewesen (v.a. um 7 Uhr morgens).

Ein herzliches Dankeschön an alle anderen Laborkollegen für die angenehme Atmosphäre, die stete Hilfsbereitschaft, die gute Laune, die schöne gemeinsame Zeit im Labor und auch die lustigen Abende. Danke also an Hang, Chris, Eddy, Ireen, Julia, und die Carreras-Crew mit Prof. Dr. Marina Kreutz, Kaste, Katrin, Eva, Alice, Gabi, Monika W., sowie Ute, Sandra und Martina.

Auch bei meinen früheren Laborkollegen Sabine, Tobi, Mike, Monika E. und Alex möchte ich mich recht herzlich bedanken. Vielen, vielen Dank, Mike, für deine Hilfe und Geduld bei meinen Computerproblemen.

Herzlichen Dank auch an Kristina und Julia für den Seelenbeistand und die Abende außerhalb des Labors and Altug für seine lustige und beruhigende Art, sowie Dagi und Nico für die lustigen Gespräche, die Gummibärchen und die in letzter Zeit stark nachlassenden McFit Besuche!

I thank Dr. Mathias Ehrich, Dr. Dirk van den Boom and Dr. Karsten Schmidt at Sequenom, Inc. San Diego for giving me the fantastic opportunity to get practical experience with the MassARRAY system and for the great time in San Diego. In particular, I would like to thank Mathias for the advice given to me at all stages of this study and Tricia Zwiefelhofer for her support, help and the training on the MassCleave technology. I would also like to thank all lab members from Sequenom for the great time and funny evenings outside the lab. I will never forget it!

Herzlichen Dank an alle im Forschungsbau H1 und allen Forschern drumherum, die ich nicht alle aufzählen kann, für den schönen Laboralltag.

Nicht zuletzt geht mein Dank an all meine Freunde, die immer für mich da waren. Danke an Flo, für deine Geduld und Unterstützung und dass du immer für mich da warst! Eoin, thanks so much for your help with the corrections!

Zum Schluss möchte ich mich noch bei meiner Familie bedanken, die immer für mich da ist. Insbesondere geht mein Dank an meine Eltern, die mir durch ihre Unterstützung und den seelischen Beistand mein Studium und diese Arbeit erst ermöglicht haben.

I want morebooks!

Buy your books fast and straightforward online - at one of world's fastest growing online book stores! Environmentally sound due to Print-on-Demand technologies.

Buy your books online at
www.morebooks.shop

Kaufen Sie Ihre Bücher schnell und unkompliziert online – auf einer der am schnellsten wachsenden Buchhandelsplattformen weltweit! Dank Print-On-Demand umwelt- und ressourcenschonend produziert.

Bücher schneller online kaufen
www.morebooks.shop

KS OmniScriptum Publishing
Brivibas gatve 197
LV-1039 Riga, Latvia
Telefax: +371 686 204 55

info@omniscriptum.com
www.omniscriptum.com

Printed by Books on Demand GmbH, Norderstedt / Germany